Pre-Algebra: The All in One Textbook

Haneul Choi

Contents

1 Exploring Whole Numbers and Integers   **13**
    Definition and Characteristics of Whole Numbers . . 13
    Definition and Characteristics of Integers 13
    Fundamental Properties and Operations 14
    Representation on the Number Line and Ordering . 14
    Multiple Choice Questions 15
    Practice Problems 17
    Answers 19

2 Place Value and the Structure of the Number System   **22**
    Understanding the Place Value System 22
    Formation of Numbers Through Digit Arrangement . 22
    The Significance of the Base-Ten Structure 23
    Multiple Choice Questions 23
    Practice Problems 26
    Answers 28

3 Basic Arithmetic Operations: Addition and Subtraction   **31**
    Fundamentals of Addition with Whole Numbers . . . 31
    Strategies and Mental Math Techniques in Addition 31
    Fundamentals of Subtraction with Whole Numbers . 32
    Strategies and Mental Math Techniques in Subtraction 32
    Connecting Addition and Subtraction
    through Arithmetic Relationships 33
    Multiple Choice Questions 33
    Practice Problems 36
    Answers 38

4 Basic Arithmetic Operations: Multiplication and Division — 42
    Fundamentals of Multiplication 42
    Techniques and Properties in Multiplication 43
    Fundamentals of Division 43
    Techniques for Efficient Division 44
    Multiple Choice Questions 45
    Practice Problems 47
    Answers 49

5 Order of Operations — 51
    Establishing a Standardized Computational Framework 51
    The Conventional Hierarchy of Operations 51
    Evaluating Complex Expressions Step by Step 52
    Strategies for Managing Multiple Operations 52
    The Role and Impact of Grouping Symbols 52
    Multiple Choice Questions 53
    Practice Problems 56
    Answers 58

6 Factors, Multiples, and Prime Numbers — 62
    Factors: Definition and Illustrative Examples 62
    Multiples: Definition and Systematic Development . 62
    Interrelations Between Factors and Multiples 63
    Prime Numbers: Core Characteristics and Fundamental Role 63
    Multiple Choice Questions 64
    Practice Problems 66
    Answers 68

7 Prime Factorization and Greatest Common Factors — 71
    Understanding Prime Factorization 71
    Methods for Breaking Numbers into Prime Factors . 72
    Determining the Greatest Common Factor Using Prime Factors 72
    Multiple Choice Questions 73
    Practice Problems 76
    Answers 77

8 Divisibility Rules — 80
- Understanding the Concept of Divisibility — 80
- Divisibility by 2, 5, and 10 — 80
- Divisibility by 3 and 9 — 81
- Divisibility by 4, 6, and 8 — 81
- Multiple Choice Questions — 81
- Practice Problems — 84
- Answers — 86

9 Understanding Fractions — 88
- Fractions as Parts of a Whole — 88
- Components of a Fraction — 88
- Visual Representations of Fractions — 89
- Basic Fraction Terminology and Concepts — 89
- Multiple Choice Questions — 89
- Practice Problems — 92
- Answers — 94

10 Equivalent Fractions and Simplification — 96
- Definition and Concept of Equivalent Fractions — 96
- Methods for Generating Equivalent Fractions — 97
- Strategies for Simplifying Fractions — 97
- Visual Representations and Conceptual Illustrations — 98
- Worked Examples and Illustrations — 98
- Multiple Choice Questions — 99
- Practice Problems — 101
- Answers — 103

11 Operations with Fractions: Addition and Subtraction — 107
- Foundational Concepts in Fraction Operations — 107
- Identifying and Establishing a Common Denominator — 107
- Addition of Fractions with Common Denominators — 108
- Subtraction of Fractions with Common Denominators — 109
- Worked Examples in Fraction Addition and Subtraction — 109
- Multiple Choice Questions — 111
- Practice Problems — 114
- Answers — 116

12 Operations with Fractions: Multiplication and Division — 118

- Multiplication of Fractions 118
  - 1 Multiplying Numerators and Denominators . 118
  - 2 Simplification through Cancellation of Common Factors 119
- Division of Fractions 119
  - 1 Reciprocals in the Division Process 119
  - 2 Performing Division by Multiplying by the Reciprocal 120
- Multiple Choice Questions 120
- Practice Problems 123
- Answers 125

13 Introduction to Decimals and Place Value — 129

- Transitioning from Fractions to Decimals 129
- Understanding Decimal Notation 129
- Place Value in a Base-Ten System 130
- Multiple Choice Questions 130
- Practice Problems 133
- Answers 135

14 Comparing, Rounding, and Ordering Decimals — 138

- Comparing Decimals 138
- Rounding Decimals 139
- Ordering Decimals 139
- Multiple Choice Questions 140
- Practice Problems 142
- Answers 144

15 Operations with Decimals: Addition and Subtraction — 147

- Decimal Place Value and Alignment in Operations . 147
- Addition of Decimals 148
  - 1 Step-by-Step Process for Adding Decimals .. 148
  - 2 Worked Example: Addition Operation 148
- Subtraction of Decimals 149
  - 1 Step-by-Step Process for Subtracting Decimals 149
  - 2 Worked Example: Subtraction Operation .. 149
- Ensuring Accuracy in Decimal Operations 150
- Multiple Choice Questions 150
- Practice Problems 153

Answers 155

16 Operations with Decimals: Multiplication and Division
158
    Multiplication of Decimals 158
    1   Procedure for Multiplying Decimals 158
    2   Worked Example: Multiplying Decimals . . . 159
    Division of Decimals 159
    1   Procedure for Dividing Decimals 159
    2   Worked Example: Dividing Decimals 159
    Multiple Choice Questions 160
    Practice Problems 163
    Answers 165

17 Understanding Percents   168
    Definition of Percents 168
    Relationship of Percents to Fractions 168
    Relationship of Percents to Decimals 169
    Conversion Techniques 169
    Examples of Conversions 170
    Multiple Choice Questions 171
    Practice Problems 174
    Answers 175

18 Percent Problems and Conversions   178
    Converting Percents to Decimals and Fractions . . . 178
    Converting Decimals and Fractions to Percents . . . 179
    Solving Percent Problems in Applied Situations . . . 179
    Worked Examples and Practice Problems 180
    Multiple Choice Questions 181
    Practice Problems 183
    Answers 185

19 Ratios and Proportions   187
    Understanding Ratios 187
    Representing and Simplifying Ratios 187
    Exploring Proportional Relationships 188
    Solving Proportions 188
    Applications of Ratios and Proportions 188
    Multiple Choice Questions 189
    Practice Problems 191
    Answers 193

20 Solving Proportional Problems — 196
- Direct Proportionality and Equivalent Ratios 196
- Establishing Proportional Equations 197
- Solving Proportions Using Cross Multiplication ... 197
- Worked Examples 197
- Applications to Problem Solving 199
- Multiple Choice Questions 199
- Practice Problems 202
- Answers 204

21 Rates and Unit Rates — 207
- Definition of Rates 207
- Understanding Unit Rates 207
- Calculating Rates and Unit Rates 208
- Interpreting Rates in Context 208
- Worked Examples 208
- Multiple Choice Questions 209
- Practice Problems 211
- Answers 213

22 Estimation and Mental Math Strategies — 215
- The Concept of Estimation 215
- Techniques for Estimation 215
  - 1 Rounding 215
  - 2 Front-End Estimation 216
  - 3 Clustering and Compatible Numbers 216
- Strategies for Mental Calculations 216
  - 1 Decomposing Numbers 216
  - 2 Utilizing the Distributive Property 217
  - 3 Simplifying Complex Operations Mentally .. 217
- Approximating Results in Arithmetic Operations .. 217
  - 1 Estimation in Addition and Subtraction ... 217
  - 2 Estimation in Multiplication and Division .. 218
- Applications in Problem Solving 218
- Multiple Choice Questions 218
- Practice Problems 221
- Answers 223

23 Introduction to Variables and Algebraic Expressions 226
- Understanding Variables 226
  - 1 Definition and Notation 226

    2   Role of Variables in Mathematical Generalization 227
Forming Algebraic Expressions 227
    1   Components of Algebraic Expressions 227
    2   Operating with Algebraic Symbols 227
Transitioning from Arithmetic to Algebra 228
    1   Generalization of Arithmetic Operations . . . 228
    2   Mathematical Modeling Using Variables . . . 228
Multiple Choice Questions 229
Practice Problems 231
Answers 233

24 Writing and Evaluating Algebraic Expressions   235

Translating Verbal Descriptions into Algebraic Expressions 235
    1   Identifying Keywords and Mathematical Operations 235
    2   Assigning Variables and Forming Expressions 235
    3   Detailed Process for Translation 236
Evaluating Algebraic Expressions for Given Values . 236
    1   Method of Substitution and Arithmetic Computation 236
    2   Worked Examples and Step-by-Step Evaluation 236
    3   Ensuring Accuracy through Verification . . . 237
Multiple Choice Questions 237
Practice Problems 239
Answers 241

25 The Distributive Property   243

Definition and Fundamental Concepts 243
    1   Definition of the Distributive Property 243
    2   Underlying Rationale 243
Expanding Algebraic Expressions 244
    1   Procedure for Expansion 244
    2   Expansion with Multiple Terms 244
Simplification Through the Distributive Property . . 245
    1   Combining Like Terms After Expansion . . . 245
    2   Reverse Application: Factoring 245
    3   Detailed Example 245
Multiple Choice Questions 246
Practice Problems 248

Answers 250

26 Combining Like Terms 253
Definition of Like Terms 253
Identifying Like Terms in Algebraic Expressions . . . 253
Process of Combining Like Terms 254
Worked Examples 254
Common Pitfalls and Considerations 255
Multiple Choice Questions 255
Practice Problems 258
Answers 260

27 Solving One-Step Equations 263
Definition and Characteristics of One-Step Equations 263
Inverse Operations and Their Application 263
Solving Equations Involving Addition and Subtraction 264
Solving Equations Involving Multiplication and Division 264
Worked Examples 265
Observations and Common Error Patterns 266
Multiple Choice Questions 267
Practice Problems 269
Answers 271

28 Solving Multi-Step Equations 274
Recognizing Multi-Step Equations 274
Combining Like Terms and Eliminating Parentheses 274
Applying Inverse Operations Sequentially 275
Worked Examples 276
Multiple Choice Questions 277
Practice Problems 280
Answers 282

29 Equations with Variables on Both Sides 286
Understanding the Structure of the Equation 286
Transposing Variable Terms 286
Combining Like Terms and Isolating the Variable . . 287
Strategies for Avoiding Common Errors 287
Worked Examples 288
Multiple Choice Questions 289
Practice Problems 291
Answers 294

30 Introduction to Inequalities — 298
- Definition and Fundamental Concepts 298
- Inequality Symbols and Their Meanings 298
- Algebraic Techniques for Solving Inequalities 299
- Representation of Solution Sets 299
- Worked Examples 299
- Multiple Choice Questions 300
- Practice Problems 303
- Answers 305

31 Solving Multi-Step Inequalities — 308
- Fundamental Concepts and Properties 308
- Procedure for Isolating the Variable 309
- Combining Like Terms and Distributing Factors . . 309
- Techniques Involving Negative Multiplication or Division 310
- Worked Example of a Multi-Step Inequality 310
- Graphical Representation and Interval Notation . . . 311
- Multiple Choice Questions 311
- Practice Problems 314
- Answers 316

32 Exploring Absolute Value — 320
- Definition and Meaning of Absolute Value 320
- Key Properties of Absolute Value 320
- Application in Equations 321
- Application in Inequalities 322
- Multiple Choice Questions 323
- Practice Problems 325
- Answers 327

33 Understanding Exponents — 330
- Definition of Exponents 330
- Exponents as Repeated Multiplication 330
- Notation and Terminology 331
- Illustrative Examples 331
- Special Cases and Fundamental Rules 332
- Applications in Mathematical Expressions 332
- Multiple Choice Questions 332
- Practice Problems 334
- Answers 336

34 Properties of Exponents — 339
- The Product Rule for Exponents 339
- The Quotient Rule for Exponents 339
- The Power Rule for Exponents 340
- Multiple Choice Questions 340
- Practice Problems 342
- Answers 344

35 Exponents with Negative and Zero Powers — 347
- Zero Exponents 347
- Negative Exponents 348
- Applications in Computations 348
- Multiple Choice Questions 349
- Practice Problems 352
- Answers 354

36 Introduction to Radicals — 357
- Definition and Notation of Radicals 357
- Square Roots as the Inverse of Exponentiation . . . 357
- Properties of Square Roots 358
- Interpretation of Square Roots in Computation . . . 359
- Multiple Choice Questions 359
- Practice Problems 362
- Answers 364

37 Simplifying Radical Expressions — 366
- Techniques for Extracting Perfect Square Factors . . 366
- Rationalizing Denominators 367
- Multiple Choice Questions 369
- Practice Problems 371
- Answers 373

38 Scientific Notation — 376
- Definition and Fundamental Principles 376
- Expressing Very Large Numbers in Compact Form . 376
- Expressing Very Small Numbers in Compact Form . 377
- Conversion Process and Detailed Examples 377
- Multiple Choice Questions 378
- Practice Problems 380
- Answers 382

39 Calculations in Scientific Notation   **385**
    Multiplication in Scientific Notation 385
    Division in Scientific Notation 386
    Addition and Subtraction in Scientific Notation . . . 386
    Normalization of Results 387
    Multiple Choice Questions 387
    Practice Problems 390
    Answers 392

40 Problem Solving Strategies in Pre-Algebra   **395**
    Approaching Mathematical Problems with Logical
    Reasoning 395
    Developing Systematic Approaches 396
    Implementing a Sequential Problem Solving Process 396
    Utilizing Estimation and Verification Techniques . . 397
    Illustrative Examples of Strategic Problem Solving . 397
    Multiple Choice Questions 398
    Practice Problems 400
    Answers 402

41 Analyzing Patterns and Sequences   **407**
    Foundations of Numeric Patterns 407
    Systematic Identification of Patterns 407
    Techniques for Extending Sequences 408
    Worked Examples and Detailed Analysis 408
    Verification of Identified Patterns 409
    Multiple Choice Questions 409
    Practice Problems 412
    Answers 414

42 Working with Formulas and Word Problems   **417**
    Understanding Mathematical Formulas in Real-Life
    Contexts 417
    Translating Real-Life Scenarios into Mathematical
    Expressions 418
    1   Identifying Key Quantities and Variables . . 418
    2   Formulating Equations from Contextual Descriptions 418
    Approaches to Solving Word Problems Through Formulas 419
    1   Analyzing the Given Problem Statement . . . 419

2	Implementing Algebraic Techniques to Solve for Unknowns	419
3	Verifying Solutions with Contextual Consistency	420

Multiple Choice Questions 420
Practice Problems 423
Answers 425

Chapter 1

Exploring Whole Numbers and Integers

Definition and Characteristics of Whole Numbers

Whole numbers comprise a set of numbers that begins with zero and continues with the succession of positive numbers. This set is represented as
$$\{0, 1, 2, 3, \ldots\},$$
and serves as a fundamental foundation in mathematics. The structure of whole numbers reflects an inherent order that allows for counting and sequential progression. Their properties include closure under addition and multiplication; that is, the sum or product of any two whole numbers results in another whole number. An intrinsic feature of whole numbers is the presence of an identity element for addition, which is the number 0. This property is essential in the development of arithmetic operations and provides the basis for more complex mathematical constructs.

Definition and Characteristics of Integers

Integers extend the concept of whole numbers by including the corresponding negative counterparts of positive whole numbers. The complete set of integers is commonly notated as

$$\mathbb{Z} = \{\ldots, -3, -2, -1, 0, 1, 2, 3, \ldots\},$$

where the addition of negatives allows for algebraic manipulation beyond simple counting. The introduction of negative values into the number system resolves the limitations inherent in whole numbers when addressing subtraction and the concept of balance between opposite quantities. Among the properties of integers are closure under addition and subtraction, commutativity for both addition and multiplication, and the presence of additive inverses for every element. These essential characteristics underpin various operations in arithmetic and algebra and provide a rigorous framework for understanding number systems.

Fundamental Properties and Operations

Both whole numbers and integers exhibit several core properties that facilitate their use in mathematical operations. One notable property is the commutative law, the principle that the order in which numbers are added or multiplied does not affect the result. Additionally, the associative property enables the grouping of numbers in addition or multiplication without altering the outcome. The distributive property interconnects addition and multiplication, stating that multiplying a sum by a number is equivalent to multiplying each addend individually by that number before summing the products. These properties not only simplify computations but also illustrate the logical structure underlying arithmetic operations. The behavior of whole numbers and integers under various operations establishes them as reliable and predictable tools for modeling numerical relationships.

Representation on the Number Line and Ordering

A number line offers a visual representation of both whole numbers and integers, emphasizing their sequential and ordered nature. On this line, whole numbers are positioned starting at zero and extend indefinitely to the right, indicating a progression of increasing magnitude. In contrast, integers include points located to the left of zero, representing negative values, and to the right, representing positive values. The even spacing between points on the number line reflects the uniform distance between consecutive whole numbers and integers. This geometric representation highlights the concept of order and magnitude, as well as the ability to compare and sequence numbers with precision. The elegant arrangement of numbers on the number line serves as a powerful tool for visualizing numerical relationships and deepening the understanding of basic arithmetic concepts.

Multiple Choice Questions

1. Which of the following sets correctly defines the whole numbers?

    (a) $\{\ldots, -3, -2, -1, 0, 1, 2, 3, \ldots\}$
    (b) $\{1, 2, 3, \ldots\}$
    (c) $\{0, 1, 2, 3, \ldots\}$
    (d) $\{0.5, 1.5, 2.5, \ldots\}$

2. Which property is true for the whole numbers when performing addition and multiplication?

    (a) They are closed under subtraction.
    (b) They are closed under addition and multiplication.
    (c) Every number has a multiplicative inverse.
    (d) They follow the non-commutative law.

3. What is the additive identity in the set of whole numbers?

    (a) 1
    (b) -1
    (c) 0
    (d) None of the above

4. How do the integers extend the set of whole numbers?

    (a) By including decimal numbers.
    (b) By including fractions.
    (c) By including the negative counterparts of whole numbers.

(d) By including irrational numbers.

5. Which property allows you to rearrange the numbers in an addition expression without changing the sum?

    (a) Associative property
    (b) Commutative property
    (c) Distributive property
    (d) Identity property

6. Which of the following best illustrates the distributive property?

    (a) a + (b + c) = (a + b) + c
    (b) a × (b + c) = a × b + a × c
    (c) a + b = b + a
    (d) a × b = b × a

7. How does a number line help in understanding whole numbers and integers?

    (a) It displays numbers at irregular intervals to show varying magnitudes.
    (b) It visually represents the sequential order with equal spacing for negative, zero, and positive values.
    (c) It is used only for whole numbers.
    (d) It represents numbers without indicating their order.

Answers:

1. **C: {0, 1, 2, 3, ... }**
   Whole numbers start at 0 and include all the positive integers in order. Option C is the correct definition.

2. **B: They are closed under addition and multiplication**
   Whole numbers have the property that adding or multiplying any two whole numbers always produces another whole number. They are not generally closed under subtraction, and not every whole number has a multiplicative inverse.

3. **C: 0**
   The additive identity is the number that, when added to any whole number, leaves the number unchanged. This is true for 0.

4. **C: By including the negative counterparts of whole numbers**
   The set of integers, denoted by $\mathbb{Z}$, extends the whole numbers by introducing the negatives of the whole numbers, which is exactly what option C states.

5. **B: Commutative property**
   The commutative property of addition states that the order of addends can be changed without affecting the sum, which is why rearranging numbers in an addition expression does not change the result.

6. **B: a × (b + c) = a × b + a × c**
   This option correctly states the distributive property, which connects multiplication with addition by distributing the multiplying factor to each term inside the parentheses.

7. **B: It visually represents the sequential order with equal spacing for negative, zero, and positive values**
   A number line is a useful visual aid that shows the ordered nature of numbers, including the positioning of whole numbers and integers with regular spacing, making it easier to compare and understand their relationships.

Practice Problems

1. Define whole numbers. In your answer, include the set notation for whole numbers and list at least two fundamental properties that they possess.

2. Explain the difference between whole numbers and integers. Provide examples in your explanation to illustrate the distinction.

3. On a number line, plot and label the following integers:
$$-4, -2, 0, 2, 4.$$

4. Evaluate the expression
$$3 + (-5)$$
and explain how the concept of additive inverses applies to this computation.

5. Provide an example that demonstrates the commutative property of addition and the associative property of multiplication using integers. Include your computations and explain why these properties hold.

6. Arrange the following integers in order from least to greatest:

   5, −3, 0, −7, 2.

Answers

1. **Answer:** Whole numbers are defined as the set of nonnegative integers. They are expressed in set notation as
   $$\{0, 1, 2, 3, \ldots\}.$$
   Two key properties of whole numbers are:
   - **Closure under Addition:** Adding any two whole numbers always produces another whole number. For example, $2 + 3 = 5$, and 5 is a whole number.
   - **Closure under Multiplication:** Multiplying any two whole numbers results in another whole number. For example, $2 \times 3 = 6$, which is also a whole number.

   These properties ensure that whole numbers form a reliable and consistent number system for basic arithmetic operations.

2. **Answer:** Whole numbers consist of numbers that start at 0 and increase without including any negative numbers; they are expressed as
   $$\{0, 1, 2, 3, \ldots\}.$$
   In contrast, integers expand on this concept by including the negatives of whole numbers. The set of integers is denoted by
   $$\mathbb{Z} = \{\ldots, -3, -2, -1, 0, 1, 2, 3, \ldots\}.$$

For instance, while 4 is a whole number, −4 is not a whole number but is considered an integer. This inclusion of negative values allows for subtraction operations that can result in numbers less than zero.

3. **Answer:** To plot the integers

$$-4, -2, 0, 2, 4$$

on a number line, begin by drawing a horizontal line. Mark a central point as 0. Then, mark equally spaced points to the right for 2 and 4, and equally spaced points to the left for −2 and −4. This diagram visually represents the ordered nature of integers: numbers increase as you move to the right and decrease toward the left.

4. **Answer:** The expression

$$3 + (-5)$$

evaluates to −2. This is because adding a negative number is equivalent to subtracting its positive value, so:

$$3 + (-5) = 3 - 5 = -2.$$

This operation demonstrates the concept of additive inverses: every number has an opposite which, when added together, sums to zero. In this example, the additive inverse of 5 is −5, so their effect on 3 results in a net decrease.

5. **Answer:**

   - **Commutative Property of Addition:** This property states that the order in which two numbers are added does not affect their sum. For example, consider:

     $$4 + (-7) = (-7) + 4.$$

     Both computations yield −3, confirming that the order of addition can be switched without affecting the result.

   - **Associative Property of Multiplication:** This property implies that when multiplying three or more numbers, the way in which the numbers are grouped does not change the product. For instance:

     $$(-2 \times 3) \times 4 = -2 \times (3 \times 4).$$

Calculating each side:

$$(-2 \times 3) \times 4 = (-6) \times 4 = -24,$$

$$-2 \times (3 \times 4) = -2 \times 12 = -24.$$

As both groupings yield -24, the associative property of multiplication is confirmed.

6. **Answer:** To arrange the integers

$$5, \ -3, \ 0, \ -7, \ 2$$

from least to greatest, identify the smallest (most negative) number and then work upward to the largest. The correct order is:
$$-7, \ -3, \ 0, \ 2, \ 5.$$

This ordering is consistent with the positions of these numbers on the number line, where values increase as you move from left (negative) to right (positive).

Chapter 2

Place Value and the Structure of the Number System

Understanding the Place Value System

In the decimal numeral system, each digit within a number is assigned a value determined by its position. The rightmost digit is associated with the ones place, which corresponds to the multiplier 10^0. Moving leftward, each subsequent digit occupies a place representing an increasing power of ten. For example, the numeral

$$582 = 5 \times 10^2 + 8 \times 10^1 + 2 \times 10^0$$

demonstrates that the digit 5 represents five hundreds, the digit 8 represents eight tens, and the digit 2 represents two ones. This structured approach to forming numbers underlines the systematic nature inherent in the place value system.

Formation of Numbers Through Digit Arrangement

The decimal system is built upon ten distinct digits: 0, 1, 2, 3, 4, 5, 6, 7, 8, and 9. When these digits are arranged in a sequence, each digit contributes to the overall number by being multiplied by the corresponding power of ten based on its position. A numeral such as

$$1,234 = 1 \times 10^3 + 2 \times 10^2 + 3 \times 10^1 + 4 \times 10^0$$

exemplifies this process. The digit in the thousands place is multiplied by 10^3, the digit in the hundreds place by 10^2, and so forth. This method of combining digits ensures that each numeral is not merely a string of symbols, but rather a precise representation of a value determined by the additive contributions of its individual components.

The Significance of the Base-Ten Structure

The decimal system is often referred to as the base-ten system because it employs ten unique digits and utilizes powers of ten to assign value to each digit's position. This base-ten structure is instrumental in simplifying arithmetic operations, such as carrying over in addition and borrowing in subtraction, by providing a uniform framework for computation. The reliance on powers of ten—10^0, 10^1, 10^2, and so on—ensures that every position within a numeral is linked to a specific magnitude. This organized structure not only facilitates accurate computations but also offers a coherent method for interpreting the size and scale of numbers. The base-ten system, with its orderly progression and clear methodological foundation, plays a central role in the representation and manipulation of numerical information.

Multiple Choice Questions

1. Which of the following best represents the value of the digit 5 in the number 582?

    (a) 5

    (b) 50

    (c) 500

    (d) 5,800

2. Which expression correctly represents the number 1,234 using the place value system?

(a) $1 \times 10^3 + 2 \times 10^2 + 3 \times 10^1 + 4 \times 10^0$

(b) $1 \times 10^2 + 2 \times 10^1 + 3 \times 10^0 + 4 \times 10^{-1}$

(c) $1 \times 10^4 + 2 \times 10^3 + 3 \times 10^2 + 4 \times 10^1$

(d) $1 \times 10^0 + 2 \times 10^1 + 3 \times 10^2 + 4 \times 10^3$

3. Why is the decimal numeral system commonly called the base-ten system?

    (a) Because it employs ten unique digits (0 through 9)
    (b) Because every number is a multiple of ten
    (c) Because it only uses the symbols 1 through 10
    (d) Because it is based on the weight of physical objects

4. Which of the following best describes how each digit in a numeral contributes to its overall value?

    (a) Each digit stands alone without the influence of its position.
    (b) Each digit is multiplied by a power of ten based on its position.
    (c) The digits are added together without any multiplication.
    (d) Only the leftmost digit determines the number's value.

5. How does an understanding of place value benefit arithmetic operations such as addition and subtraction?

    (a) It allows students to ignore the significance of digit positions.
    (b) It provides a framework to align numbers according to their magnitudes.
    (c) It eliminates the need for memorizing arithmetic facts.
    (d) It simplifies computation by only focusing on the largest digits.

6. In the numeral 582, which power of ten is associated with the digit 2?

    (a) 10^0
    (b) 10^1

(c) 10^2

   (d) 10^3

7. Why is the base-ten system considered fundamental in mathematics?

   (a) Because it minimizes the use of large numbers.

   (b) Because it introduces a consistent method for interpreting digits based on their positional value.

   (c) Because it ignores the role of zero in number representation.

   (d) Because it is solely used in theoretical arithmetic.

Answers:

1. **C: 500**
   Explanation: In the number 582, the digit 5 is in the hundreds place. This means it represents 5×10^2, which equals 500.

2. **A: $1 \times 10^3 + 2 \times 10^2 + 3 \times 10^1 + 4 \times 10^0$**
   Explanation: The numeral 1,234 is correctly expressed by multiplying each digit by its corresponding power of ten. Here, 1 is in the thousands place (10^3), 2 in the hundreds place (10^2), 3 in the tens place (10^1), and 4 in the ones place (10^0).

3. **A: Because it employs ten unique digits (0 through 9)**
   Explanation: The decimal system is called the base-ten system because it uses ten distinct symbols (0 through 9) to form all numbers.

4. **B: Each digit is multiplied by a power of ten based on its position**
   Explanation: In the place value system, every digit's value is determined by its position in the numeral, which dictates the power of ten by which the digit is multiplied.

5. **B: It provides a framework to align numbers according to their magnitudes**
   Explanation: Understanding place value allows students to correctly align digits by their respective powers of ten, which is essential for operations such as carrying in addition and borrowing in subtraction.

6. **A: 10^0**
   Explanation: In the number 582, the digit 2 is located in the ones place, which corresponds to 10^0 (or 1). Hence, it represents 2 ones.

7. **B: Because it introduces a consistent method for interpreting digits based on their positional value**
   Explanation: The base-ten system is fundamental because it provides an organized structure where each digit's value is clearly defined by its position (through powers of ten), simplifying both the reading of numbers and performing arithmetic operations.

Practice Problems

1. Write the expanded form of the number

    $$4375$$

    and explain the role of each digit based on its place value.

2. In the number

    $$1246$$

    identify the place value of the digit 2 and explain how its position determines its value.

3. Write the numeral corresponding to the expanded form:

$$6 \times 10^3 + 3 \times 10^2 + 5 \times 10^1 + 8 \times 10^0$$

and explain your process of converting it to standard form.

4. Explain why the decimal numeral system is referred to as the base-ten system. Include in your explanation the significance of having ten unique digits.

5. Arrange the following numbers in ascending order:

$$307, \quad 370, \quad 703$$

and describe how the concept of place value helps determine their order.

6. Given the number
$$482$$
swap the digit in the hundreds place with the digit in the tens place to form a new number. Explain how this change in the positions of the digits affects the overall value of the number.

Answers

1. **Solution:** To write the number 4375 in expanded form, break it down by each digit's place value:
$$4375 = 4 \times 10^3 + 3 \times 10^2 + 7 \times 10^1 + 5 \times 10^0.$$

   Here, the digit 4 is in the thousands place and represents 4000, the digit 3 is in the hundreds place representing 300, 7 is in the tens place representing 70, and 5 in the ones place represents 5. Each digit is multiplied by a corresponding power of ten based on its position, which is the key idea behind the place value system.

2. **Solution:** In the number 1246, the digits correspond to:
$$1 \times 10^3, \quad 2 \times 10^2, \quad 4 \times 10^1, \quad 6 \times 10^0.$$

   The digit 2 is in the hundreds place, which means it represents:
$$2 \times 10^2 = 200.$$

   This shows that the position of a digit in a number determines its value by associating it with a specific power of ten.

3. **Solution:** The given expanded form is
$$6 \times 10^3 + 3 \times 10^2 + 5 \times 10^1 + 8 \times 10^0.$$
To convert it into a numeral:
$$6 \times 10^3 = 6000, \quad 3 \times 10^2 = 300, \quad 5 \times 10^1 = 50, \quad 8 \times 10^0 = 8.$$
Adding these together:
$$6000 + 300 + 50 + 8 = 6358.$$
Therefore, the numeral is 6358. This process illustrates how each term in the expanded form contributes to the total value based on its place value.

4. **Solution:** The decimal numeral system is called the base-ten system because it uses ten unique digits: 0, 1, 2, 3, 4, 5, 6, 7, 8, and 9. In this system, each digit's position represents a power of ten (such as 10^0, 10^1, 10^2, etc.), which makes arithmetic operations like addition, subtraction, carrying, and borrowing more systematic and clear. The use of ten digits is historically linked to the fact that humans typically have ten fingers, providing a natural counting base.

5. **Solution:** When arranging the numbers 307, 370, and 703 in ascending order, compare their digits based on place value. Both 307 and 370 have 3 in the hundreds place. However, for 307, the tens digit is 0 and the ones digit is 7, while for 370, the tens digit is 7 and the ones digit is 0, making 307 smaller than 370. The number 703 begins with a 7 in the hundreds place, which makes it the largest. Thus, the numbers in ascending order are:
$$307, \quad 370, \quad 703.$$
This ordering is determined by comparing each positional value from the highest (hundreds) to the lowest (ones).

6. **Solution:** In the number 482, the original place values are:
$$4 \times 10^2 + 8 \times 10^1 + 2 \times 10^0.$$
If we swap the digit in the hundreds place with the digit in the tens place, the number becomes:
$$8 \times 10^2 + 4 \times 10^1 + 2 \times 10^0.$$

Calculating each term gives:

$$8 \times 10^2 = 800, \quad 4 \times 10^1 = 40, \quad 2 \times 10^0 = 2.$$

Adding these values together:

$$800 + 40 + 2 = 842.$$

This change shows how switching positions alters the value of each digit by changing which power of ten it is multiplied by. The hundreds digit, when swapped, now represents a higher or lower value depending on its new position.

Chapter 3

Basic Arithmetic Operations: Addition and Subtraction

Fundamentals of Addition with Whole Numbers

Addition is the process of combining two or more whole numbers to produce a sum. Each digit is aligned according to its place value, ensuring that hundreds, tens, and ones are properly accounted for during calculation. For example, in the problem

$$457 + 289,$$

the numbers are arranged so that the hundreds, tens, and ones are added in their respective columns. The ones digits combine, then the tens, and finally the hundreds, resulting in a coherent structure that supports systematic computation.

Strategies and Mental Math Techniques in Addition

Various techniques are available to streamline addition and facilitate mental computation. One effective strategy involves decomposing numbers into parts that represent their place value compo-

nents. Taking the numbers 457 and 289, these can be separated into:
$$457 = 400 + 50 + 7 \quad \text{and} \quad 289 = 200 + 80 + 9.$$

By adding the corresponding components,
$$400 + 200 = 600, \quad 50 + 80 = 130, \quad \text{and} \quad 7 + 9 = 16,$$

and then combining the intermediate sums,
$$600 + 130 + 16 = 746,$$

the overall sum is determined in a clear and methodical manner. Techniques such as rounding to the nearest ten and counting on also play a role in enhancing mental calculation skills. Aligning numbers by their place value remains central to ensuring that all contributions are properly recognized and combined.

Fundamentals of Subtraction with Whole Numbers

Subtraction identifies the difference between whole numbers as the inverse process of addition. In the subtraction problem
$$832 - 457,$$

the digits are organized according to the ones, tens, and hundreds places. When a digit in the minuend is smaller than the corresponding digit in the subtrahend, the method of borrowing (regrouping) is employed. This approach guarantees that each subtraction step respects the hierarchical structure dictated by the place value system, thereby maintaining accuracy throughout the calculation.

Strategies and Mental Math Techniques in Subtraction

The use of mental math extends to subtraction as well, where breaking a number into its constituents simplifies the reduction process. For instance, the number 832 may be thought of as 800, 30, and 2, while 457 can be viewed as 400, 50, and 7. Subtraction is then carried out for each corresponding place value, with adjustments made whenever regrouping is necessary. A typical process might involve initially subtracting the hundreds,

$$800 - 400 = 400,$$

handling the tens by temporarily compensating for insufficient value,

$$30 - 50 \quad \text{(with regrouping)},$$

and finally subtracting the ones. This decomposition not only reveals the structure behind subtraction but also builds a flexible approach to solving problems using mental techniques. Methods such as rounding and then compensating for the rounding adjustment further enhance the ability to perform subtractions quickly while maintaining precision.

Connecting Addition and Subtraction through Arithmetic Relationships

A fundamental connection exists between addition and subtraction based on their inverse relationship. The equation

$$a + b = c$$

implies that
$$c - a = b \quad \text{and} \quad c - b = a.$$

This arithmetic symmetry reinforces the understanding of how numbers relate to one another within the whole number system. The clear organization of digits by place value supports both operations, allowing the use of mental math techniques to check the accuracy of calculations. Recognizing that subtraction undoes addition bolsters numerical fluency and lays the groundwork for flexible problem-solving strategies in arithmetic.

Multiple Choice Questions

1. When adding whole numbers like 457 and 289, why is it important to align the numbers by their place values?

    (a) To ensure that digits in the same place (ones, tens, hundreds, etc.) are added together correctly.

(b) To mix the digits for a faster calculation.

   (c) To convert the numbers into decimals.

   (d) To simplify the process by ignoring the digits in the ones place.

2. Which of the following shows the best way to decompose the number 457 for easier addition?

   (a) 400 + 50 + 7

   (b) 450 + 7

   (c) 457 cannot be decomposed.

   (d) 300 + 100 + 57

3. When subtracting 457 from 832, borrowing (regrouping) is necessary because:

   (a) Some digits in the minuend are smaller than the corresponding digits in the subtrahend.

   (b) Subtraction always begins with the hundreds place.

   (c) It converts subtraction into addition.

   (d) Borrowing makes the calculation more complex.

4. Which mental math strategy can help simplify subtraction problems?

   (a) Rounding numbers to the nearest ten, subtracting, and then adjusting the result.

   (b) Ignoring the ones and tens places.

   (c) Multiplying the digits before subtracting.

   (d) Reversing the order of the digits.

5. How are addition and subtraction connected through their arithmetic relationship?

   (a) Subtraction undoes addition.

   (b) Subtraction is the same as multiplication.

   (c) They are unrelated operations.

   (d) Addition is the inverse of division.

6. Decomposing numbers into their place value components (for example, writing 289 as 200 + 80 + 9) is a useful strategy because:

(a) It simplifies mental computation by allowing you to add or subtract each part separately.

(b) It makes calculations more difficult by creating extra steps.

(c) It is only useful for addition, not subtraction.

(d) It eliminates the need to align digits by place value.

7. Why is it important to work with digits according to their place values when performing addition or subtraction?

(a) It ensures that ones, tens, and hundreds are handled correctly during the calculation.

(b) It allows the problem to be converted into a multiplication problem.

(c) It permits you to ignore smaller digits.

(d) It simplifies the calculation by using only the largest digits in each number.

Answers:

1. **A: To ensure that digits in the same place (ones, tens, hundreds, etc.) are added together correctly.**
Aligning the numbers by place value is fundamental in addition because it guarantees that each column's digits (ones with ones, tens with tens, etc.) are combined accurately.

2. **A: 400 + 50 + 7**
Breaking 457 into 400, 50, and 7 reflects its place value components, making it easier to add by addressing each part separately.

3. **A: Some digits in the minuend are smaller than the corresponding digits in the subtrahend.**
Borrowing, or regrouping, is needed in subtraction when a digit in the minuend (the number you are subtracting from) is too small to subtract the corresponding digit of the subtrahend.

4. **A: Rounding numbers to the nearest ten, subtracting, and then adjusting the result.**
This mental math strategy simplifies the subtraction process by dealing with round numbers first and then compensating for any rounding discrepancies.

5. **A: Subtraction undoes addition.**
   The inverse relationship between addition and subtraction means that if a + b = c, then c - a = b. Recognizing this connection reinforces understanding of both operations.

6. **A: It simplifies mental computation by allowing you to add or subtract each part separately.**
   Decomposing numbers based on place value makes calculations more manageable, as it enables students to work with simpler, smaller numbers instead of one large number.

7. **A: It ensures that ones, tens, and hundreds are handled correctly during the calculation.**
   Working according to place value maintains the integrity of the arithmetic process, ensuring that each digit contributes correctly to the overall sum or difference.

Practice Problems

1. Evaluate the following addition problem:
$$457 + 289$$

2. Use a decomposition strategy to compute the sum:
$$324 + 178$$

3. Solve the subtraction problem using regrouping:
$$832 - 457$$

4. Solve the subtraction problem by applying the regrouping method:
$$1002 - 586$$

5. Write the addition equation that corresponds to the subtraction:

Given that
$$832 - 457 = 375,$$

write an addition equation that confirms the inverse relationship between addition and subtraction.

6. Using mental math and the decomposition method, evaluate the sum:
$$635 + 274$$

Answers

1. Evaluate the following addition problem:
$$457 + 289 = 746.$$

   Solution: First, align the numbers by their place values:

   Ones Place: $7 + 9 = 16$. Write down 6 and carry over 1 to the tens place.

   Tens Place: $5 + 8 + 1 = 14$. Write down 4 and carry over 1 to the hundreds place.

   Hundreds Place: $4 + 2 + 1 = 7$.

   Combining these results gives the final sum:
   $$746.$$

2. Use a decomposition strategy to compute the sum:
$$324 + 178.$$

   Solution: Break each number into its place value components:
   $$324 = 300 + 20 + 4,$$
   $$178 = 100 + 70 + 8.$$

Now, add the corresponding components:

$$300 + 100 = 400,$$
$$20 + 70 = 90,$$
$$4 + 8 = 12.$$

Finally, combine the sums:

$$400 + 90 + 12 = 502.$$

Therefore, the sum is 502.

3. Solve the subtraction problem using regrouping:

$$832 - 457.$$

Solution: Write the numbers with their digits aligned by place value. Since the ones digit of 832 (which is 2) is less than the ones digit of 457 (which is 7), we need to regroup.

Step 1 (Ones Place): Borrow 1 ten from the tens digit of 832. The ones digit becomes 12:

$$12 - 7 = 5.$$

Step 2 (Tens Place): After borrowing, the tens digit of 832 reduces from 3 to 2. Now subtract:

$$2 - 5.$$

Since 2 is less than 5, borrow 1 hundred from the hundreds digit. The tens digit becomes 12 (after borrowing):

$$12 - 5 = 7.$$

Step 3 (Hundreds Place): The hundreds digit of 832, after borrowing, becomes 7:

$$7 - 4 = 3.$$

Thus, the final answer is:

$$375.$$

4. Solve the subtraction problem by applying the regrouping method:
$$1002 - 586.$$

Solution: Write the number 1002 in its place values:

$$1 \quad 0 \quad 0 \quad 2$$

and the number 586 as:

$$0 \quad 5 \quad 8 \quad 6$$

Step 1 (Ones Place): Since $2 < 6$, we cannot subtract directly. The tens digit is 0, so we move to the hundreds place. Borrow 1 from the thousands place:

$$1 \to 0 \quad \text{and} \quad 0 + 10 = 10 \text{ (hundreds digit)}.$$

Step 2: Now, borrow 1 from the hundreds (10 becomes 9) for the tens place, making the tens digit $0 + 10 = 10$. Then, borrow 1 from the tens (10 becomes 9) to add to the ones place:

$$\text{Ones digit: } 2 + 10 = 12.$$

Now perform the subtraction:

$$\text{Ones: } 12 - 6 = 6,$$
$$\text{Tens: } 9 - 8 = 1,$$
$$\text{Hundreds: } 9 - 5 = 4,$$
$$\text{Thousands: } 0 - 0 = 0.$$

Putting the digits together, we obtain:

$$416.$$

5. Write the addition equation that corresponds to the subtraction:

$$832 - 457 = 375.$$

Solution: The inverse relationship between addition and subtraction means that if subtracting 457 from 832 gives 375, then adding 457 and 375 should return 832. Thus, the corresponding addition equation is:

$$457 + 375 = 832.$$

This verifies the connection between the two operations.

6. Using mental math and the decomposition method, evaluate the sum:
$$635 + 274.$$

Solution: First, break each number into its place value parts:

$$635 = 600 + 30 + 5,$$
$$274 = 200 + 70 + 4.$$

Now, add the corresponding components:

$$600 + 200 = 800,$$
$$30 + 70 = 100,$$
$$5 + 4 = 9.$$

Finally, combine the sums:

$$800 + 100 + 9 = 909.$$

Therefore, the sum is 909.

Chapter 4

Basic Arithmetic Operations: Multiplication and Division

Fundamentals of Multiplication

Multiplication is defined as the process of combining equal groups and is equivalent to repeated addition. For instance, the expression
$$3 \times 4 = 4 + 4 + 4 = 12$$
demonstrates that multiplying 3 by 4 produces a sum equivalent to adding four three times. The structure of multiplication is supported by several important properties. The commutative property establishes that the order of factors does not impact the product, as shown in
$$a \times b = b \times a.$$
The associative property explains that when several factors are involved, the grouping of numbers is irrelevant:
$$(a \times b) \times c = a \times (b \times c).$$
Furthermore, the identity property states that any number multiplied by one remains unchanged:
$$a \times 1 = a.$$

These properties underlie the systematic computation methods used in multiplication and are pivotal in strategies that simplify more complex problems.

Techniques and Properties in Multiplication

Efficient computation in multiplication is achieved through various techniques that streamline the process. A common method involves decomposing numbers into their place value components. For example, when calculating

$$23 \times 15,$$

one may represent 23 as

$$23 = 20 + 3.$$

Then, the multiplication can be broken into parts as follows:

$$20 \times 15 = 300 \quad \text{and} \quad 3 \times 15 = 45.$$

The partial products are then combined:

$$300 + 45 = 345.$$

The distributive property, expressed by

$$a \times (b + c) = a \times b + a \times c,$$

provides the theoretical basis for this partitioning strategy. In addition, visual models such as the area model or array representations display how each component of the factors contributes to the overall product. Such techniques afford clarity in the multiplication process, ensuring that every contribution from each place value is accurately accounted for before summing partial results.

Fundamentals of Division

Division serves as the inverse operation of multiplication and is understood as the process of partitioning a number into a specified

number of equal parts. The operation is often symbolized by the division sign ($\div$) or expressed in fraction form as
$$\frac{a}{b}.$$
For example, the equation
$$56 \div 7 = 8$$
indicates that 56 consists of eight groups of 7. In algebraic terms, if a is the dividend and b is the divisor, then there exists a quotient q and a remainder r satisfying
$$a = b \times q + r,$$
with the condition that $0 \leq r < b$. Unlike multiplication, division does not possess the commutative property; in general,
$$a \div b \neq b \div a,$$
except in cases where the numbers in question are identical. This fundamental asymmetry necessitates careful consideration when performing division operations.

Techniques for Efficient Division

Division can be executed efficiently through systematic methods such as long division, estimation, and the decomposition of the dividend. The long division algorithm begins by determining how many times the divisor fits into the leading portion of the dividend. For example, consider the division
$$987 \div 3.$$
The process involves dividing the leftmost group of digits by 3, writing the corresponding quotient digit, multiplying the divisor by that digit, subtracting the product from the current portion of the dividend, and then bringing down the next digit. This cycle continues until the entire dividend has been processed. Additionally, estimation techniques rely on a strong understanding of multiplication facts and help in predicting the quotient digit at each step. An alternative strategy is to decompose the dividend into parts that are easier to divide, thereby partitioning the division process into simpler, manageable segments. Methods that combine estimation with the standard algorithm enable efficient computation, ensuring that each step reduces the potential for error while providing a clear and logical approach to division.

Multiple Choice Questions

1. Which of the following best describes multiplication?

    (a) A process of repeated subtraction

    (b) A process of combining equal groups, which is equivalent to repeated addition

    (c) A process of counting numbers in order

    (d) A process of finding differences between numbers

2. Which property of multiplication states that changing the order of the numbers does not change the product?

    (a) Distributive property

    (b) Commutative property

    (c) Associative property

    (d) Identity property

3. When multiplying 23 by 15 using decomposition, which of the following is the best approach?

    (a) Express 23 as 20 + 3, multiply each by 15, and then add the results

    (b) Break 15 into 10 + 5, multiply each part by 23, and then subtract the results

    (c) Divide 23 into two equal groups and then multiply by 15

    (d) Add 23 and 15 first, then multiply the sum by a common factor

4. Division is best described as the inverse operation of which process?

    (a) Addition

    (b) Subtraction

    (c) Multiplication

    (d) Exponentiation

5. In long division, what is the correct initial step when solving $987 \div 3$?

(a) Determine how many times 3 fits into the first digit (9) of the dividend

(b) Subtract 3 from 987 repeatedly until reaching zero

(c) Combine the last two digits of the dividend and divide by 3

(d) Estimate the full quotient directly without breaking down the dividend

6. The equation
$$a \times (b + c) = a \times b + a \times c$$
demonstrates which property of multiplication?

(a) Commutative property

(b) Associative property

(c) Identity property

(d) Distributive property

7. Which of the following statements about division is false?

(a) Division is the inverse operation of multiplication.

(b) Division can result in a quotient with a remainder.

(c) Division is commutative, meaning that a ÷ b equals b ÷ a.

(d) Estimation and decomposition strategies can improve the efficiency of division.

Answers:

1. **B:** Multiplication is defined as the process of combining equal groups, effectively adding the same number repeatedly. This is why 3 × 4 can be seen as 4 + 4 + 4.

2. **B:** The commutative property of multiplication states that the order of the factors does not affect the product, which means a × b is the same as b × a.

3. **A:** Decomposing 23 into 20 and 3 allows you to multiply each piece by 15 separately. Using the distributive property, you then add the products (20 × 15 and 3 × 15) to obtain the final answer.

4. **C:** Division is the inverse of multiplication. This means that performing division undoes multiplication, separating a total into equal parts.

5. **A:** In the long division method, you start by determining how many times the divisor (3) fits into the leftmost part of the dividend (9 in 987), establishing the first digit of the quotient.

6. **D:** The equation a × (b + c) = a × b + a × c shows the distributive property. This property is essential in breaking down complex multiplication problems into simpler, manageable parts.

7. **C:** Division does not have the commutative property; that is, in general, a ÷ b is not equal to b ÷ a. This distinguishes division from multiplication, which is commutative.

Practice Problems

1. Calculate the product:

$$23 \times 15$$

    using the decomposition method (i.e. express 23 as the sum of its place value components) and then combine the partial products.

2. Verify the commutative and associative properties of multiplication for the numbers 4, 5, and 6. Provide two examples—one for each property.

3. Using the long division algorithm, compute:
$$987 \div 3$$
Show all the steps of your division.

4. Solve the division problem with a remainder:
$$59 \div 7$$
Identify both the quotient and the remainder.

5. Explain why division is not commutative by comparing the results of:
$$48 \div 6 \quad \text{and} \quad 6 \div 48.$$
Provide a clear explanation.

6. Word Problem: A baker has 36 cupcakes. If these cupcakes are divided equally among 4 trays, and then each tray is further divided into 3 equal groups for display purposes, determine the number of cupcakes in each group.

Answers

1. To compute 23×15 using the decomposition method, begin by breaking 23 into its place value components:
$$23 = 20 + 3$$
Multiply each component by 15:
$$20 \times 15 = 300 \quad \text{and} \quad 3 \times 15 = 45$$
Finally, add the partial products:
$$300 + 45 = 345$$
Therefore,
$$23 \times 15 = 345$$

2. The commutative property of multiplication tells us that the order of factors does not change the product. For example:
$$4 \times 5 = 5 \times 4 = 20$$
The associative property indicates that when three numbers are multiplied, the way in which they are grouped does not affect the product. For instance,
$$(4 \times 5) \times 6 = 20 \times 6 = 120,$$
while
$$4 \times (5 \times 6) = 4 \times 30 = 120$$
Both examples confirm the properties.

3. To perform the long division of 987 ÷ 3, proceed as follows:
   - Divide the first digit: 3 goes into 9 exactly 3 times (since $3 \times 3 = 9$). Subtract 9 from 9, leaving 0.
   - Bring down the next digit (8). Divide 8 by 3: 3 fits into 8 two times ($3 \times 2 = 6$) with a remainder of 2 (since $8 - 6 = 2$).
   - Bring down the final digit (7) to join the remainder, forming 27. Divide 27 by 3: 3 goes into 27 exactly 9 times ($3 \times 9 = 27$).

   Thus, the quotient is 329 with no remainder.

4. For 59 ÷ 7:
   - Determine how many times 7 fits into 59. Since $7 \times 8 = 56$ and $7 \times 9 = 63$ (which is too large), the quotient is 8.
   - Compute the remainder by subtracting 56 from 59:
   $$59 - 56 = 3$$

   Therefore, the division yields a quotient of 8 and a remainder of 3.

5. Division is not commutative because changing the order of the dividend and divisor changes the outcome. For example,
   $$48 \div 6 = 8,$$
   whereas
   $$6 \div 48 = \frac{6}{48} = \frac{1}{8}$$
   Since 8 is not equal to $\frac{1}{8}$, this demonstrates that, unlike multiplication, the order in division matters.

6. In the word problem, start by dividing 36 cupcakes equally among 4 trays:
   $$36 \div 4 = 9 \quad \text{(cupcakes per tray)}$$
   Next, divide each tray into 3 equal groups:
   $$9 \div 3 = 3 \quad \text{(cupcakes per group)}$$
   Therefore, each group contains 3 cupcakes.

Chapter 5

Order of Operations

Establishing a Standardized Computational Framework

Arithmetic expressions that combine several operations require a systematic process to yield unambiguous results. In such expressions, the potential for differing interpretations is eliminated by adopting a conventional hierarchy that specifies which operations are performed first. This standardized framework ensures consistency in calculations regardless of the complexity of the expression.

The Conventional Hierarchy of Operations

The established sequence for evaluating expressions begins with operations enclosed within grouping symbols. Parentheses, brackets, and braces indicate that the expression contained within them must be resolved prior to addressing any external operations. Once the grouping symbols have been evaluated, the next step is to compute any exponentiation present in the expression. Multiplication and division follow, executed from left to right since they share the same level of precedence. The final operations to be performed are addition and subtraction, also processed from left to right. This hierarchy, often remembered by the mnemonic "Parentheses, Exponents, Multiplication and Division, Addition and Subtraction," provides a clear and definitive method for approaching even the most intricate arithmetic expressions.

Evaluating Complex Expressions Step by Step

Consider the expression

$$(3+5) \times 2^2 - \frac{12}{3+1}.$$

The evaluation proceeds as follows. First, the arithmetic within the grouping symbols is calculated: in the first pair of parentheses, $3 + 5$ is computed to obtain 8; in the denominator of the fraction, $3 + 1$ equals 4. Next, the exponent is addressed; 2^2 is evaluated, resulting in 4. With these intermediate values determined, the multiplication is performed, yielding $8 \times 4 = 32$, and the division is carried out to compute $\frac{12}{4} = 3$. Finally, the subtraction $32 - 3$ is executed, resulting in 29. Each step observes the hierarchy of operations, showcasing how the systematic approach guarantees that every operation is applied in the correct order.

Strategies for Managing Multiple Operations

Clarity in the evaluation of expressions with multiple operations is achieved by rewriting or restructuring the expression to emphasize its grouping. Introducing additional grouping symbols can further highlight the intended calculation order. When multiplication and division or addition and subtraction occur in sequence, the operations must be performed from left to right due to their equal precedence. Careful and deliberate organization of the steps involved in the computation minimizes errors and reinforces precision. This methodical approach allows each operation to be verified before proceeding to the next, ensuring that the overall evaluation remains both accurate and consistent.

The Role and Impact of Grouping Symbols

Grouping symbols are fundamental in controlling the order in which operations are executed. Their strategic placement directs the evaluator to resolve specific parts of an expression before applying the general hierarchy. By enclosing certain components within parentheses, brackets, or braces, it becomes possible to override the default sequence of operations and prioritize particular calculations. This capability is especially useful in constructing complex expressions that require clarity in their structure. The deliberate use of grouping symbols not only clarifies the intended order of operations but also enhances the overall precision and readability of mathematical expressions.

Multiple Choice Questions

1. Which of the following best represents the standard hierarchy of operations (often remembered by the mnemonic PEMDAS)?

    (a) Parentheses, Multiplication and Division, Exponents, Addition and Subtraction

    (b) Parentheses, Exponents, Multiplication and Division, Addition and Subtraction

    (c) Exponents, Parentheses, Addition and Subtraction, Multiplication and Division

    (d) Parentheses, Addition and Subtraction, Exponents, Multiplication and Division

2. When evaluating a complex arithmetic expression, which operation should always be performed first?

    (a) Exponentiation

    (b) Multiplication

    (c) Operations inside grouping symbols

    (d) Addition

3. When an expression includes both multiplication and division, how are these operations handled?

    (a) Always perform multiplication before division.

    (b) Always perform division before multiplication.

    (c) Evaluate them from right to left.

    (d) Evaluate them from left to right.

4. Consider the expression:
$$(2+6) \div 2^2 + 3.$$

Which sequence of steps correctly follows the order of operations?

   (a) Compute $2+6$, then 2^2, followed by division, and finally addition.
   (b) Compute 2^2, then $2+6$, followed by division, and finally addition.
   (c) Add 3 to $2+6$ first, then compute 2^2 and finally perform division.
   (d) Compute $2+6$ and 2^2 simultaneously, then add 3 before performing division.

5. Which of the following is true regarding the use of grouping symbols (parentheses, brackets, or braces) in an expression?

   (a) They can be evaluated at any time as long as the final answer is correct.
   (b) They have the lowest priority in the order of operations.
   (c) They ensure that the expression inside them is calculated before addressing any outside operations.
   (d) They are used solely for improving the visual clarity of the expression.

6. The mnemonic PEMDAS is used to recall the order in which operations are performed. What does each letter in PEMDAS stand for?

   (a) Parentheses, Exponents, Multiplication and Division, Addition and Subtraction.
   (b) Parentheses, Exponents, Multiplication, Division, Addition, Subtraction.
   (c) Parentheses, Multiplication, Exponents, Division, Addition, Subtraction.
   (d) Parentheses, Exponents, Division and Multiplication, Addition and Subtraction.

7. In an expression that involves only addition and subtraction, such as
$$5 - 3 + 2,$$
what is the proper method to evaluate it?

   (a) Always perform the addition first, then the subtraction.

   (b) Always perform the subtraction first, then the addition.

   (c) Evaluate the operations from left to right.

   (d) Evaluate the operations from right to left.

Answers:

1. **B: Parentheses, Exponents, Multiplication and Division, Addition and Subtraction**
   This option correctly reflects the well-established order of operations. Parentheses are handled first, followed by exponentiation, then multiplication and division (from left to right), and finally addition and subtraction (from left to right).

2. **C: Operations inside grouping symbols**
   Grouping symbols (parentheses, brackets, or braces) indicate that the expression within them must be evaluated first, which helps to avoid ambiguity in complex expressions.

3. **D: Evaluate them from left to right**
   Multiplication and division have the same level of precedence. When they occur in sequence, they are performed from left to right as they appear in the expression.

4. **A: Compute $2 + 6$, then 2^2, followed by division, and finally addition**
   The correct procedure is to first resolve the grouping symbol by computing 2+6, then handle exponentiation by computing 2^2, perform the division next, and add 3 at the end.

5. **C: They ensure that the expression inside them is calculated before addressing any outside operations**
   Grouping symbols have the highest priority in the order of operations, which means the expression within them is always evaluated first to clarify the intended computation.

6. **A: Parentheses, Exponents, Multiplication and Division, Addition and Subtraction**

PEMDAS stands for Parentheses, Exponents, Multiplication and Division, Addition and Subtraction. This mnemonic helps students remember the proper sequence of operations to avoid misinterpretation.

7. **C: Evaluate the operations from left to right**
In cases where only addition and subtraction are present, the operations are performed sequentially from left to right according to the standard order of operations.

Practice Problems

1. Evaluate the expression:
$$(3+2) \times 4 - 6 \div 2$$

2. Evaluate the expression:
$$8 \div 2(2+2)$$

3. Evaluate the expression:
$$3 + 6 \times (5 + 4) \div 3 - 7$$

4. Evaluate the expression:
$$(12 - (3 \times 2)^2) \div 3 + 1$$

5. Evaluate the expression:
$$2^3 \times (4 - 2) + 10 \div 2$$

6. Evaluate the expression:
$$9 - 3^2 + (6 \div 2)^2$$

Answers

1. For the expression:
$$(3 + 2) \times 4 - 6 \div 2$$
   Solution:

   (a) First, evaluate the grouping inside the parentheses:
   $$3 + 2 = 5.$$

   (b) Next, perform the multiplication:
   $$5 \times 4 = 20.$$

   (c) Then, compute the division:
   $$6 \div 2 = 3.$$

   (d) Finally, perform the subtraction:
   $$20 - 3 = 17.$$

   Therefore, the value of the expression is 17.

2. For the expression:
$$8 \div 2(2 + 2)$$
   Solution:

(a) First, evaluate the grouping:
$$2 + 2 = 4.$$

(b) The expression becomes:
$$8 \div 2 \times 4.$$

(c) According to the order of operations, perform division and multiplication from left to right. First divide:
$$8 \div 2 = 4.$$

(d) Then, multiply:
$$4 \times 4 = 16.$$

Thus, the expression evaluates to 16. (Note: Although some might interpret the expression differently, following the left-to-right rule for division and multiplication yields 16.)

3. For the expression:
$$3 + 6 \times (5 + 4) \div 3 - 7$$

Solution:

(a) Start by evaluating the grouping:
$$5 + 4 = 9.$$

(b) Substitute to get:
$$3 + 6 \times 9 \div 3 - 7.$$

(c) Next, perform the multiplication:
$$6 \times 9 = 54.$$

(d) Follow with the division:
$$54 \div 3 = 18.$$

(e) Finally, perform the addition and subtraction from left to right:
$$3 + 18 = 21 \quad \text{and} \quad 21 - 7 = 14.$$

Hence, the final answer is 14.

4. For the expression:
$$(12 - (3 \times 2)^2) \div 3 + 1$$

   Solution:

   (a) First, compute the inner multiplication:
   $$3 \times 2 = 6.$$

   (b) Next, apply the exponent:
   $$6^2 = 36.$$

   (c) Evaluate the outer parentheses:
   $$12 - 36 = -24.$$

   (d) Then, perform the division:
   $$-24 \div 3 = -8.$$

   (e) Finally, add:
   $$-8 + 1 = -7.$$

   Therefore, the expression evaluates to -7.

5. For the expression:
$$2^3 \times (4 - 2) + 10 \div 2$$

   Solution:

   (a) First, evaluate the exponent:
   $$2^3 = 8.$$

   (b) Next, evaluate the grouping:
   $$4 - 2 = 2.$$

   (c) Multiply the results:
   $$8 \times 2 = 16.$$

(d) Evaluate the division:
$$10 \div 2 = 5.$$

(e) Finally, add:
$$16 + 5 = 21.$$

Thus, the value of the expression is 21.

6. For the expression:
$$9 - 3^2 + (6 \div 2)^2$$

Solution:

(a) First, compute the exponent in the middle of the expression:
$$3^2 = 9.$$

(b) Next, evaluate the grouping:
$$6 \div 2 = 3,$$
then apply the exponent:
$$3^2 = 9.$$

(c) Substitute back into the expression to get:
$$9 - 9 + 9.$$

(d) Perform the operations in order:
$$9 - 9 = 0,$$
and then:
$$0 + 9 = 9.$$

Therefore, the final answer is 9.

Chapter 6

Factors, Multiples, and Prime Numbers

Factors: Definition and Illustrative Examples

A factor of a whole number is defined as any number that divides the given number evenly, leaving no remainder. In other words, if an integer a divides an integer b exactly, then a is considered a factor of b. For instance, the number 12 can be expressed as a product of several paired factors:

$$12 = 1 \times 12, \quad 12 = 2 \times 6, \quad 12 = 3 \times 4.$$

The complete list of factors for 12 includes the numbers 1, 2, 3, 4, 6, and 12. This method of expressing a number as products of factors provides a detailed look at its internal structure and lays the groundwork for understanding more complex numerical interactions.

Multiples: Definition and Systematic Development

A multiple of a number is obtained by multiplying that number by an integer. For any given integer n, the sequence of multiples is generated by the expression

$$n, 2n, 3n, 4n, \ldots,$$

which continues indefinitely. For example, taking the number 3, its multiples are produced as follows:

$$3, 6, 9, 12, 15, \ldots$$

This concept not only reinforces the idea of repeated addition but also demonstrates the scalability of numbers. The systematic progression of multiples reveals recurring patterns that are key to understanding arithmetic operations and the relationships between different numbers.

Interrelations Between Factors and Multiples

The concepts of factors and multiples are intertwined in the study of numbers. When examining two distinct numbers, the common factors serve to identify shared divisors, while common multiples highlight the regular intervals at which these numbers coincide when multiplied by integers. Consider the numbers 12 and 18. The factors of 12 are 1, 2, 3, 4, 6, and 12, and the factors of 18 are 1, 2, 3, 6, 9, and 18. The set of common factors for these two numbers is 1, 2, 3, and 6. Such comparisons are useful in various applications such as finding the greatest common factor (GCF) or the least common multiple (LCM), tools that facilitate the analysis of ratios and the simplification of fractions.

Prime Numbers: Core Characteristics and Fundamental Role

Prime numbers constitute a special category within the collection of whole numbers. A prime number is characterized by having exactly two distinct positive factors: 1 and the number itself. For example, the numbers 2, 3, 5, 7, and 11 meet this criterion because none of them can be divided evenly by any number besides 1 and the number itself. The restriction inherent in the definition of prime numbers makes them the indivisible building blocks of composite numbers. This idea is encapsulated in the representation of any composite number as a product of prime factors:

$$n = p_1 \times p_2 \times \cdots \times p_k,$$

where each p_i is a prime number. The unique factorization of composite numbers into primes serves as a fundamental principle in number theory, illuminating the ways in which numbers interact and combine. Prime numbers thus occupy a central role in the understanding of mathematical structure and arithmetic processes.

Multiple Choice Questions

1. Which of the following best defines a factor?
    (a) A number that is the result of multiplying two numbers
    (b) A number that divides another number evenly, with no remainder
    (c) A number that is only divisible by 1 and itself
    (d) A number that is added repeatedly to form another number

2. All of the following are factors of 12 except:
    (a) 1
    (b) 3
    (c) 5
    (d) 4

3. Which of the following statements correctly describes a multiple of a number?
    (a) It is a number that can only be divided by 1 and itself.
    (b) It is a number that divides evenly into the given number.
    (c) It is obtained by multiplying the given number by an integer.
    (d) It is the result of adding the given number to itself repeatedly.

4. What is the 5th multiple of 3?
    (a) 9

(b) 12

(c) 15

(d) 18

5. For the numbers 12 and 18, which set represents their common factors?

    (a) 1, 2, 3, 6
    (b) 1, 2, 4
    (c) 1, 3, 12
    (d) 1, 2, 6, 12, 18

6. Which of the following numbers is prime?

    (a) 9
    (b) 15
    (c) 17
    (d) 21

7. Which statement best explains the role of prime numbers in the factorization of composite numbers?

    (a) Every composite number can be uniquely expressed as a product of prime numbers.
    (b) Composite numbers are formed by adding prime numbers together.
    (c) Prime numbers are the only factors found in composite numbers.
    (d) Every multiple of a composite number must include a prime number.

Answers:

1. **B: A number that divides another number evenly, with no remainder**
   A factor is defined as any number that divides another number without leaving a remainder.

2. **C: 5**
   The factors of 12 are 1, 2, 3, 4, 6, and 12. Since 5 does not divide 12 evenly, it is not a factor.

3. **C: It is obtained by multiplying the given number by an integer**
   Multiples are generated by multiplying a number by an integer, which produces a sequence such as n, 2n, 3n, etc.

4. **C: 15**
   The 5th multiple of 3 is calculated as $3 \times 5 = 15$.

5. **A: 1, 2, 3, 6**
   The common factors of 12 (1, 2, 3, 4, 6, 12) and 18 (1, 2, 3, 6, 9, 18) are the numbers that evenly divide both, namely 1, 2, 3, and 6.

6. **C: 17**
   A prime number has exactly two factors: 1 and itself. Among the options, 17 meets this criterion.

7. **A: Every composite number can be uniquely expressed as a product of prime numbers**
   This is a statement of the Fundamental Theorem of Arithmetic, which tells us that the prime factorization of any composite number is unique (apart from the order of the factors). This principle makes primes the building blocks of all composite numbers.

Practice Problems

1. Consider the number 24. List all of its factors and explain why each number divides 24 evenly.

2. Find and list the first eight multiples of 5. In your explanation, describe how these multiples are generated and why the sequence continues indefinitely.

3. Determine the greatest common factor (GCF) and the least common multiple (LCM) of 18 and 24. Show your work and explain each step of your reasoning.

4. Explain why the number 1 is not considered a prime number. Use the definition of a prime number in your explanation.

5. Determine whether the number 29 is a prime number. Provide a step-by-step explanation of your reasoning.

6. Express the composite number 60 as a product of its prime factors by using the factor tree method. Provide a clear explanation or diagram of your process.

Answers

1. **Solution:** To list the factors of 24, we search for all whole numbers that divide 24 evenly (i.e., with no remainder). The factor pairs for 24 are:

$$1 \times 24, \quad 2 \times 12, \quad 3 \times 8, \quad 4 \times 6.$$

   Thus, the complete list of factors is:

$$1, 2, 3, 4, 6, 8, 12, 24.$$

   Each of these numbers divides 24 exactly, which is why they are all factors of 24.

2. **Solution:** Multiples of a number are found by multiplying that number by the integers 1, 2, 3, and so on. For the number 5, the first eight multiples are:

$$5 \times 1 = 5, \quad 5 \times 2 = 10, \quad 5 \times 3 = 15, \quad 5 \times 4 = 20,$$
$$5 \times 5 = 25, \quad 5 \times 6 = 30, \quad 5 \times 7 = 35, \quad 5 \times 8 = 40.$$

   These multiples (5, 10, 15, 20, 25, 30, 35, and 40) are generated by repeated addition (or multiplication) of 5. The sequence continues indefinitely because there is no largest integer to multiply by.

3. **Solution:** To find the GCF and LCM of 18 and 24, start by listing the factors of each number.

   For 18, the factors are:

$$1, 2, 3, 6, 9, 18.$$

   For 24, the factors are:

$$1, 2, 3, 4, 6, 8, 12, 24.$$

   The common factors are:

$$1, 2, 3, 6.$$

   Hence, the greatest common factor (GCF) is 6.

   To find the LCM, list some multiples of each number:

Multiples of 18:

$$18, 36, 54, 72, 90, \ldots$$

Multiples of 24:

$$24, 48, 72, 96, \ldots$$

The smallest common multiple is 72. Thus, the least common multiple (LCM) is 72.

Another method uses the relationship:

$$\text{LCM} = \frac{18 \times 24}{\text{GCF}} = \frac{432}{6} = 72.$$

4. **Solution:** By definition, a prime number is a whole number that has exactly two distinct positive factors: 1 and the number itself. The number 1 only has one factor (itself), so it does not meet the criteria for having two distinct factors. Therefore, 1 is not considered a prime number.

5. **Solution:** To determine if 29 is a prime number, we check whether it has any factors other than 1 and 29. The possible divisors to test are the prime numbers less than or equal to the square root of 29. Since the square root of 29 is approximately 5.38, we only need to test the prime numbers 2, 3, and 5.

   - 29 divided by 2 yields 14.5 (not an integer). - 29 divided by 3 yields approximately 9.67 (not an integer). - 29 divided by 5 yields 5.8 (not an integer).

   Since none of these divisions produces an integer, 29 has no divisors other than 1 and itself. Thus, 29 is a prime number.

6. **Solution:** To express 60 as a product of its prime factors, we use the factor tree method:

   1. Start with 60. Break it into two factors, for example:

   $$60 = 2 \times 30.$$

   2. Next, break down 30:

   $$30 = 2 \times 15.$$

   3. Then, break down 15:

   $$15 = 3 \times 5.$$

4. At this point, 2, 3, and 5 are all prime numbers.

Therefore, the prime factorization of 60 is:
$$60 = 2 \times 2 \times 3 \times 5.$$

This can also be written as:
$$60 = 2^2 \times 3 \times 5.$$

The factor tree method helps break down the composite number step by step until every factor is prime.

Chapter 7

Prime Factorization and Greatest Common Factors

Understanding Prime Factorization

Prime factorization is the process of expressing a composite number as a product of prime numbers. A prime number is defined as a number greater than 1 that has no divisors other than 1 and itself. In contrast, a composite number has additional divisors and can be broken down into prime components. The fundamental theorem of arithmetic guarantees that every composite number has a unique prime factorization, apart from the order of the factors. For example, the number 60 can be described as a composite number that, when factored into primes, yields a representation in the form

$$60 = 2 \times 2 \times 3 \times 5,$$

which can be written using exponents as

$$60 = 2^2 \times 3 \times 5.$$

This unique representation is a powerful tool for understanding the internal structure of numbers.

Methods for Breaking Numbers into Prime Factors

One of the most common techniques for obtaining the prime factorization of a number is the factor tree method. This method involves iteratively breaking down a composite number by dividing it by the smallest prime factor available until only prime numbers remain. In this procedure, a composite number is first separated into two factors; one of these factors is chosen as the smallest prime factor that divides the number evenly. The non-prime factor, if composite, is then further divided in a similar manner. For instance, consider the number 60. The process begins by choosing the smallest prime factor, resulting in the split

$$60 = 2 \times 30.$$

Continuing the process with 30, the number is factored as

$$30 = 2 \times 15.$$

Since 15 is composite, it can be expressed further as

$$15 = 3 \times 5.$$

Collecting all the prime factors together produces the full prime factorization

$$60 = 2 \times 2 \times 3 \times 5 = 2^2 \times 3 \times 5.$$

This systematic approach using factor trees not only simplifies large numbers into their prime components but also reinforces the understanding of division and multiplication relationships among numbers.

Determining the Greatest Common Factor Using Prime Factors

The greatest common factor (GCF) of two or more numbers is the largest number that divides each of them without leaving any remainder. By utilizing prime factorization, the GCF can be determined by comparing the prime factors of the numbers involved. Each number is first expressed in its prime factorized form. Then, the common prime factors are identified and, for each common prime, the smallest exponent present in the factorizations is selected. Multiplying these common primes, raised to their respective smallest exponents, yields the GCF.

As an illustrative example, consider the numbers 24 and 36. The prime factorization of 24 is obtained by noting that

$$24 = 2 \times 2 \times 2 \times 3 = 2^3 \times 3,$$

while 36 can be broken down into

$$36 = 2 \times 2 \times 3 \times 3 = 2^2 \times 3^2.$$

The common prime factors include 2 and 3. For the prime number 2, the smallest exponent appearing in the factorizations is 2 (since the exponents are 3 for 24 and 2 for 36). For the prime number 3, the smallest exponent is 1 (with exponents 1 for 24 and 2 for 36). Consequently, the greatest common factor is calculated as

$$\text{GCF} = 2^2 \times 3 = 4 \times 3 = 12.$$

This method provides an efficient and clear procedure for determining the GCF by leveraging the unique representation of numbers through prime factorization.

Multiple Choice Questions

1. Which of the following correctly defines a prime number?

    (a) A number that has exactly two distinct positive divisors: 1 and itself

    (b) A number that has more than two divisors

    (c) A number that is only divisible by 1

    (d) A number that can be expressed as a product of two or more numbers

2. Which of these numbers is NOT a composite number?

    (a) 18

    (b) 20

    (c) 29

    (d) 30

3. What is the prime factorization of 60?

    (a) $2 \times 3 \times 10$
    (b) $2^2 \times 3 \times 5$
    (c) $2 \times 2 \times 3 \times 6$
    (d) $2^3 \times 3 \times 5$

4. In the factor tree method for finding a prime factorization, what is the first step one should take?

    (a) Write the number as a product of two prime numbers.
    (b) Divide the number by its largest factor.
    (c) Divide the number by its smallest prime factor that divides it evenly.
    (d) Add the digits of the number to determine a possible factor.

5. How is the Greatest Common Factor (GCF) of two numbers determined using their prime factorizations?

    (a) Multiply all prime factors from both numbers together.
    (b) Multiply the common prime factors raised to their highest exponents found in each number.
    (c) Multiply the common prime factors raised to their lowest exponents found in each number.
    (d) Add the common prime factors together.

6. Given the prime factorizations of 24 and 36 as $24 = 2^3 \times 3$ and $36 = 2^2 \times 3^2$, what is their GCF?

    (a) $2^2 \times 3 = 12$
    (b) $2^3 \times 3 = 24$
    (c) $2^2 \times 3^2 = 36$
    (d) $2 \times 3 = 6$

7. Which statement best describes the Fundamental Theorem of Arithmetic?

    (a) Every composite number can be uniquely expressed as a product of prime numbers, disregarding the order of the factors.

(b) Only even numbers have a unique prime factorization.

(c) Prime numbers can be factored into primes in several different ways.

(d) Composite numbers always have multiple prime factorizations.

Answers:

1. **A: A number that has exactly two distinct positive divisors: 1 and itself**
   This is the definition of a prime number. A prime number is greater than 1 and only divisible by 1 and itself, which distinguishes it from composite numbers.

2. **C: 29**
   The number 29 is prime because its only divisors are 1 and 29. Composite numbers have more than two positive divisors.

3. **B:** $2^2 \times 3 \times 5$
   The prime factorization of 60 is obtained by repeatedly dividing by the smallest prime factors: $60 = 2 \times 30$, $30 = 2 \times 15$, and $15 = 3 \times 5$. This gives $60 = 2 \times 2 \times 3 \times 5$, which can be written as $2^2 \times 3 \times 5$.

4. **C: Divide the number by its smallest prime factor that divides it evenly**
   The factor tree method starts by dividing the composite number by the smallest prime factor. This step is repeated with the quotient until all remaining factors are prime.

5. **C: Multiply the common prime factors raised to their lowest exponents found in each number**
   When using prime factorizations to find the GCF, you take each common prime and raise it to the smallest power found in the factorization of the given numbers. Multiplying these together gives the GCF.

6. **A:** $2^2 \times 3 = 12$
   For the prime factor 2, the smallest exponent between 24 (2^3) and 36 (2^2) is 2. For the prime factor 3, the smallest exponent is 1 (present in 24 as 3^1). Thus, the GCF is $2^2 \times 3 = 12$.

7. **A: Every composite number can be uniquely expressed as a product of prime numbers, disregarding the order of the factors**

The Fundamental Theorem of Arithmetic states that every integer greater than 1 is either a prime itself or can be factored into primes uniquely (except for the order in which the primes are written).

Practice Problems

1. Find the prime factorization of 56 using the factor tree method. Write your answer in exponent form.

2. Express 90 as a product of prime numbers. Use a factor tree if needed and rewrite your answer using exponents.

3. Use the factor tree method to find the prime factorization of 84. Write your final answer in exponent form.

4. Determine the greatest common factor (GCF) of the numbers 48 and 180 by first expressing each number in its prime factorized form.

5. For the numbers 72 and 108, write their prime factorizations and then determine their GCF.

6. Find the greatest common factor (GCF) of the numbers 84, 126, and 210 by using prime factorization.

Answers

1. **Solution:**
   We begin by breaking 56 into factors using a factor tree. One

way to start is:
$$56 = 7 \times 8,$$
where 7 is a prime number. Next, factor 8:
$$8 = 2 \times 4,$$
and then break 4 into:
$$4 = 2 \times 2.$$
Thus, the full prime factorization is:
$$56 = 2 \times 2 \times 2 \times 7 = 2^3 \times 7.$$

2. **Solution:**
Factorize 90 by starting with:
$$90 = 2 \times 45.$$
Next, break down 45:
$$45 = 3 \times 15,$$
and then factor 15:
$$15 = 3 \times 5.$$
Collecting all the prime factors yields:
$$90 = 2 \times 3 \times 3 \times 5 = 2 \times 3^2 \times 5.$$

3. **Solution:**
To factorize 84, start with:
$$84 = 2 \times 42.$$
Then, factor 42 as:
$$42 = 2 \times 21.$$
Finally, factor 21:
$$21 = 3 \times 7.$$
Thus, the prime factorization of 84 is:
$$84 = 2 \times 2 \times 3 \times 7 = 2^2 \times 3 \times 7.$$

4. **Solution:**
First, express 48 and 180 as products of their prime factors.
For 48:
$$48 = 2 \times 24 = 2 \times 2 \times 12 = 2 \times 2 \times 2 \times 6 = 2 \times 2 \times 2 \times 2 \times 3 = 2^4 \times 3.$$
For 180:
$$180 = 2 \times 90 = 2 \times 2 \times 45 = 2^2 \times (3 \times 15) = 2^2 \times 3 \times (3 \times 5) = 2^2 \times 3^2 \times 5.$$

Identify the common prime factors using the smallest exponents: the prime 2 appears with exponents 4 and 2 (take 2), and the prime 3 appears with exponents 1 and 2 (take 1). Therefore, the GCF is:
$$\text{GCF} = 2^2 \times 3 = 4 \times 3 = 12.$$

5. **Solution:**
First, factorize both numbers. For 72:
$$72 = 2 \times 36 = 2 \times 2 \times 18 = 2 \times 2 \times 2 \times 9 = 2^3 \times 3 \times 3 = 2^3 \times 3^2.$$
For 108:
$$108 = 2 \times 54 = 2 \times 2 \times 27 = 2^2 \times (3 \times 9) = 2^2 \times 3 \times 3 \times 3 = 2^2 \times 3^3.$$

Take the common primes with their smallest exponents: for 2, the smaller exponent is 2; for 3, it is 2. Thus, the GCF is:
$$\text{GCF} = 2^2 \times 3^2 = 4 \times 9 = 36.$$

6. **Solution:**
Determine the prime factorization of each number:
For 84:
$$84 = 2^2 \times 3 \times 7.$$
For 126:
$$126 = 2 \times 63 = 2 \times 3 \times 21 = 2 \times 3 \times 3 \times 7 = 2 \times 3^2 \times 7.$$
For 210:
$$210 = 2 \times 105 = 2 \times 3 \times 35 = 2 \times 3 \times 5 \times 7.$$

Identify the common prime factors among the three numbers with their smallest exponents: the prime 2 is present as 2^2, 2^1, and 2^1 (take 2^1); the prime 3 appears as 3^1, 3^2, and 3^1 (take 3^1); and the prime 7 appears with exponent 1 in each case. Thus, the GCF is:
$$\text{GCF} = 2^1 \times 3^1 \times 7^1 = 2 \times 3 \times 7 = 42.$$

Chapter 8

Divisibility Rules

Understanding the Concept of Divisibility

Divisibility describes a property of integers that indicates when one number can be divided by another without leaving a remainder. In mathematical terms, an integer A is divisible by an integer B if there exists an integer C such that $A = B \times C$. This fundamental idea provides a framework for simplifying calculations and breaking numbers into their basic building blocks. The use of simple divisibility rules offers a rapid method for determining factors, an essential skill in mental arithmetic and efficient problem solving.

Divisibility by 2, 5, and 10

A number is divisible by 2 when its last digit is even; that is, when the units digit is 0, 2, 4, 6, or 8. This rule emerges directly from the structure of the base-ten numbering system.

A similar straightforward rule applies for divisibility by 5. A number is divisible by 5 if its final digit is either 0 or 5. In the case of 10, a number is divisible by 10 when its last digit is 0. These rules simplify the process of identifying factors in larger numbers. For example, in evaluating a multi-digit number, a quick glance at the units digit can immediately reveal whether 2, 5, or 10 divides the number evenly.

Divisibility by 3 and 9

The rules for divisibility by 3 and 9 rely on the sum of the digits of the number in question. For divisibility by 3, if the sum of all digits results in a number that is divisible by 3, then the original number is also divisible by 3. The same pattern applies to 9: if the sum of the digits is divisible by 9, then the entire number must be divisible by 9. This method reduces a potentially complex division process to simple addition and a quick check of the resulting sum. For instance, a number whose digits add up to 18 will be divisible by both 3 and 9, since 18 is divisible by both of these numbers.

Divisibility by 4, 6, and 8

The rule for divisibility by 4 involves focusing on the last two digits of a number. If the two-digit number formed by the final two digits is divisible by 4, then the entire number is divisible by 4. This test capitalizes on the properties of multiples of 4 within the decimal system.

Divisibility by 6 requires that a number meets the criteria for both divisibility by 2 and by 3. That is, the number must end in an even digit and the sum of its digits must be divisible by 3. This combination of conditions provides a quick verification process when determining whether a number is a multiple of 6.

For divisibility by 8, the rule examines the last three digits of the number. If these three digits form a number that is divisible by 8, then the entire number is divisible by 8. This criterion is particularly useful with larger numbers for which checking the complete number may not be as efficient. The method of isolating the final three digits reduces the task to evaluating a much smaller number, thereby supporting rapid mental computation.

Each of these rules is derived from the fundamental structure of the base-ten system and offers a means to quickly identify factors. The clarity and simplicity of these tests make them a valuable tool in both academic settings and everyday arithmetic.

Multiple Choice Questions

1. Which of the following best describes the concept of divisibility?

(a) An integer A is divisible by an integer B if there exists an integer C such that A = B × C.

(b) An integer A is divisible by an integer B if A divided by B produces a decimal.

(c) An integer A is divisible by an integer B if the sum of A's digits is divisible by B.

(d) An integer A is divisible by an integer B if B is a factor of A only when A is prime.

2. Which rule correctly determines if a number is divisible by 2?

   (a) The number is divisible by 2 if its last digit is even (0, 2, 4, 6, or 8).

   (b) The number is divisible by 2 if the sum of its digits is even.

   (c) The number is divisible by 2 if its first digit is even.

   (d) The number is divisible by 2 if the product of its digits is even.

3. A number is divisible by 5 if:

   (a) Its last digit is either 1 or 5.

   (b) Its last digit is either 0 or 5.

   (c) The sum of its digits is 5 or a multiple of 5.

   (d) Its first and last digits are both 5.

4. To determine if a number is divisible by 3 or 9, which method is used?

   (a) Checking if the last digit is 3 or 9.

   (b) Adding all the digits of the number and checking if the sum is divisible by 3 (or 9).

   (c) Multiplying the digits together and seeing if the product is 3 or 9.

   (d) Comparing the first and last digits of the number.

5. How can you quickly check whether a number is divisible by 4?

   (a) By verifying if its last digit is 4.

- (b) By adding the last two digits and checking if the sum is 4.
- (c) By checking if the two-digit number formed by its last two digits is divisible by 4.
- (d) By doubling the last digit and comparing it with the first digit.

6. A number is divisible by 6 if:

    - (a) It is divisible by both 2 and 3.
    - (b) It is divisible by both 3 and 4.
    - (c) It ends with a 6.
    - (d) The sum of its digits is 6 or a multiple of 6.

7. What is the proper rule for testing divisibility by 8?

    - (a) The sum of its digits must be divisible by 8.
    - (b) Its last digit must be 8.
    - (c) Its last two digits must form a number divisible by 8.
    - (d) The number formed by its last three digits must be divisible by 8.

Answers:

1. **A:** An integer A is divisible by an integer B if there exists an integer C such that $A = B \times C$. This is the formal definition of divisibility. It indicates that the division results in an integer without a remainder.

2. **A:** The number is divisible by 2 if its last digit is even (0, 2, 4, 6, or 8). The base-ten system makes it easy to check divisibility by 2 simply by inspecting the units digit.

3. **B:** A number is divisible by 5 if its last digit is either 0 or 5. This rule is straightforward because only numbers ending in 0 or 5 can be evenly divided by 5.

4. **B:** To check divisibility by 3 or 9, add all the digits of the number. If the sum is divisible by 3, then the number is divisible by 3; similarly, if the sum is divisible by 9, then the number is divisible by 9.

5. **C:** A number is divisible by 4 if the two-digit number formed by its last two digits is divisible by 4. This method takes advantage of properties in the base-ten system, making it unnecessary to consider the entire number.

6. **A:** A number is divisible by 6 if it is divisible by both 2 and 3. Since 6 is the product of 2 and 3, both conditions must be met for a number to be a multiple of 6.

7. **D:** A number is divisible by 8 if the number formed by its last three digits is divisible by 8. Checking only the last three digits simplifies the process, especially for large numbers, and accurately determines divisibility by 8.

Practice Problems

1. Determine whether the number 358 is divisible by 2, 5, and 10. For each case, explain how the divisibility rule applies.

2. Determine whether the number 273 is divisible by 3 and by 9. Use the rule involving the sum of the digits in your explanation.

3. Check if the number 15264 is divisible by 4 and by 8. Explain your reasoning by considering the last two digits for divisibility by 4 and the last three digits for divisibility by 8.

4. Using appropriate divisibility rules, determine whether the number 114 is divisible by 6. Explain which rules you used and why.

5. Determine whether the number 725 is divisible by 2, 5, and 10. Explain your reasoning based on the final digit of the number.

6. Explain why the divisibility rule for 8 involves checking the last three digits of a number. Then, using this rule, deter-

mine whether 12304 is divisible by 8. Provide a complete explanation.

Answers

1. **Solution:** To decide the divisibility of 358 by each number, we apply the corresponding rules:

   ### Divisible by 2:

   A number is divisible by 2 if its final digit is even. Since 358 ends in 8, which is even, 358 **is divisible by 2**.

   ### Divisible by 5:

   A number is divisible by 5 if its final digit is 0 or 5. Here, the last digit is 8, so 358 **is not divisible by 5**.

   ### Divisible by 10:

   A number is divisible by 10 if it ends in 0. Since 358 ends in 8, it **is not divisible by 10**.

2. **Solution:** The divisibility rules for 3 and 9 are based on the sum of the digits.

   $$\text{Sum of digits of } 273 = 2 + 7 + 3 = 12.$$

   For divisibility by 3, since 12 is divisible by 3 ($12 \div 3 = 4$), 273 **is divisible by 3**. For divisibility by 9, the rule requires the digit sum to be divisible by 9. Since 12 is not divisible by 9, 273 **is not divisible by 9**.

3. **Solution:** To test divisibility by 4, we look at the last two digits:
$$\text{Last two digits of } 15264 = 64.$$
Since $64 \div 4 = 16$ exactly, 15264 **is divisible by 4**. To test divisibility by 8, we examine the last three digits:
$$\text{Last three digits of } 15264 = 264.$$
Because $264 \div 8 = 33$ with no remainder, 15264 **is divisible by 8**.

4. **Solution:** A number is divisible by 6 if it is divisible by both 2 and 3.
   Divisibility by 2: 114 ends in 4
   (an even digit), so it is divisible by 2.
   Divisibility by 3: Sum of digits of $114 = 1 + 1 + 4 = 6$.
   Since 6 is divisible by 3 ($6 \div 3 = 2$), 114 **is divisible by 3**. Because both conditions are met, 114 **is divisible by 6**.

5. **Solution:** For divisibility by 2, the rule requires the final digit to be even.

   725 ends in 5 (an odd number), so it is not divisible by 2.

   For divisibility by 5, a number must end in 0 or 5. Since 725 ends in 5, it **is divisible by 5**. For divisibility by 10, the number must end in 0. As 725 ends in 5, it **is not divisible by 10**.

6. **Solution:** The divisibility rule for 8 involves checking the last three digits of a number because 1000 is a multiple of 8. Any number can be expressed as a sum of a multiple of 1000 and the number formed by its last three digits. Since the multiple of 1000 is always divisible by 8, only the last three digits determine the divisibility by 8.
   For 12304, we consider the last three digits:
   $$\text{Last three digits} = 304.$$
   Dividing 304 by 8:
   $$304 \div 8 = 38,$$
   which is an exact division (no remainder). Therefore, 12304 **is divisible by 8**.

Chapter 9

Understanding Fractions

Fractions as Parts of a Whole

Fractions represent a means to express a quantity as a portion of a complete entity. In this perspective, any whole object or set is divided into a number of equal parts. The idea of fractions is rooted in the observation that a whole can be partitioned into segments that each contribute to the overall total. The concept emphasizes that a fraction is not merely a division, but rather a representation of how many parts of a given size are present compared to the number of parts that would make up the entire thing.

Components of a Fraction

Every fraction is composed of two distinct elements separated by a horizontal line. The element positioned above the line is known as the numerator, and it indicates the number of parts under consideration. The element found below the line is the denominator, which specifies the total number of equal parts that the whole is divided into. For example, in a fraction represented as a/b, the numerator (a) shows how many parts are taken, while the denominator (b) shows into how many equal parts the whole is split. This clear division of roles underpins the meaning of the fraction and aids in performing various fraction operations accurately.

Visual Representations of Fractions

Visual models play an essential role in providing an intuitive grasp of fractional values. One typical representation involves a circle or a rectangle that is partitioned into equally sized regions. In such a diagram, shading a number of these regions illustrates the value of a fraction. This method allows for the establishment of an immediate connection between the abstract numeric expression and the tangible idea of dividing a whole into equal segments. Moreover, these visual models support reasoning about the magnitude of different fractions and assist in comparing fractions by examining the relative proportions of the shaded areas.

Basic Fraction Terminology and Concepts

When discussing fractions, several key terms and concepts appear repeatedly in explanations and problem solving. The numerator is acknowledged as the counting number for parts that are selected or examined, whereas the denominator provides the framework by indicating how many parts in all are needed to form the whole. The concept of equivalent fractions arises when different pairs of numerators and denominators ultimately represent the same part-to-whole relationship after proper simplification. The notion of the simplest form of a fraction involves reducing the fraction so that the numerator and denominator share no common factors other than one. The clarity provided by consistent terminology and careful explanation enhances the understanding of fractions as an integral part of elementary arithmetic and prepares the groundwork for later topics involving fraction operations and comparisons.

Multiple Choice Questions

1. Which of the following best describes what a fraction represents?

    (a) A ratio comparing two different quantities

    (b) A part of a whole expressed as a division into equal parts

    (c) A method to perform subtraction

    (d) A representation of a whole number only

2. What are the two main components of any fraction?

    (a) Sum and difference
    (b) Numerator and denominator
    (c) Dividend and divisor
    (d) Factor and multiple

3. Which component of a fraction indicates the total number of equal parts that make up the whole?

    (a) Numerator
    (b) Denominator
    (c) Both numerator and denominator together
    (d) Neither; it is determined by the fraction's value

4. Visual models, such as a circle or rectangle divided into regions with some parts shaded, help students understand that:

    (a) Fractions are only useful for drawing pictures
    (b) A whole can be divided into equal parts where the shaded parts illustrate the fraction's value
    (c) Only the numerator matters in a fraction
    (d) The shape of the visual has no connection to the fraction's value

5. Two fractions are considered equivalent if:

    (a) They have the same numerator only
    (b) Their numerators and denominators are identical
    (c) They represent the same part-to-whole relationship even if the numbers differ
    (d) They have different values but similar appearances

6. Simplifying a fraction means:

    (a) Changing its value so that the numerator is larger than the denominator
    (b) Converting the fraction into a decimal form
    (c) Reducing it so that the numerator and denominator share no common factors besides 1
    (d) Reordering the numerator and denominator

7. In a diagram, if a rectangle is divided into 8 equal parts and 3 parts are shaded, which fraction best represents the shaded area?

   (a) 3/5
   (b) 3/8
   (c) 5/8
   (d) 8/3

Answers:

1. **B: A part of a whole expressed as a division into equal parts** This is the key idea behind fractions. A fraction represents how many parts of a given size make up the whole.

2. **B: Numerator and denominator** Every fraction consists of two components: the numerator (top number) and the denominator (bottom number).

3. **B: Denominator** The denominator tells us into how many equal parts the whole is divided, which is essential in defining the fraction.

4. **B: A whole can be divided into equal parts where the shaded parts illustrate the fraction's value** Visual models help students see the relationship between the shaded (selected) parts and the whole, reinforcing the meaning of fractions.

5. **C: They represent the same part-to-whole relationship even if the numbers differ** Equivalent fractions may look different, but they represent the same proportion of the whole after appropriate simplification.

6. **C: Reducing it so that the numerator and denominator share no common factors besides 1** Simplifying a fraction means dividing both the numerator and denominator by their greatest common factor so that the fraction is in its simplest form.

7. **B: 3/8** Since the rectangle is divided into 8 equal parts and 3 are shaded, the shaded portion of the whole is represented by the fraction 3/8.

Practice Problems

1. Identify the numerator and denominator in the fraction
$$\frac{3}{7}$$

2. Using a visual model, draw a diagram representing the fraction
$$\frac{4}{9}$$
by dividing a rectangle (or circle) into 9 equal parts and shading 4 parts.

3. Simplify the fraction
$$\frac{6}{8}$$
to its simplest form and explain your simplification process.

4. Determine whether the fractions

$$\frac{2}{3} \text{ and } \frac{4}{6}$$

are equivalent. Provide a detailed explanation.

5. A pizza is divided into 8 equal slices. If 3 slices are eaten, represent the portion eaten as a fraction and explain what it represents.

6. Explain in your own words why fractions are important for representing parts of a whole. Include a real-life example in your explanation.

Answers

1. For the fraction
$$\frac{3}{7}$$
   Solution:
   In this fraction, the numerator is 3 and the denominator is 7. The numerator (3) indicates the number of parts being considered, while the denominator (7) shows that the whole has been divided into 7 equal parts. This means that if you divide a whole into 7 equal sections, you are focusing on 3 of those sections.

2. For the fraction
$$\frac{4}{9}$$
   Solution:
   A visual model for this fraction can be created by drawing a rectangle (or a circle) and dividing it into 9 equal sections. Then, shade 4 of these sections. This diagram clearly shows that out of the 9 equal parts that make up the whole, 4 parts are being highlighted. Visualizing fractions in this way helps to bridge the gap between abstract numerical representation and a tangible understanding of part-to-whole relationships.

3. To simplify the fraction
$$\frac{6}{8}$$
   Solution:
   Begin by finding the greatest common factor (GCF) of the numerator and the denominator. The number 6 and 8 both have a common factor of 2. Dividing both the numerator and denominator by 2 gives:
$$\frac{6 \div 2}{8 \div 2} = \frac{3}{4}.$$
   The fraction $\frac{3}{4}$ is in its simplest form because 3 and 4 have no common factors other than 1.

4. To determine if
$$\frac{2}{3} \text{ and } \frac{4}{6}$$
   are equivalent, **Solution:**
   Simplify $\frac{4}{6}$ by finding the greatest common factor of 4 and

6, which is 2. Dividing the numerator and denominator by 2 yields:
$$\frac{4 \div 2}{6 \div 2} = \frac{2}{3}.$$
Since $\frac{4}{6}$ simplifies exactly to $\frac{2}{3}$, both fractions represent the same portion of a whole, meaning they are equivalent.

5. In the pizza example, the pizza is divided into 8 equal slices and 3 slices are eaten, so the fraction representing the eaten portion is
$$\frac{3}{8}.$$
   Solution:
   Here, the numerator 3 represents the number of slices that have been eaten, while the denominator 8 represents the total number of equal slices that make up the pizza. Thus, $\frac{3}{8}$ indicates that 3 out of 8 slices (or a little less than half of the pizza) have been consumed.

6. Fractions are vital for representing parts of a whole in many everyday situations. **Solution:**
   They allow us to clearly express how a whole is divided into equal parts. For example, when following a recipe, you might need $\frac{1}{2}$ cup of milk. This fraction tells you that the milk should be measured as half of a standard cup, ensuring accuracy in the recipe. Another example is sharing a chocolate bar. If the bar is divided into 10 equal pieces and you eat 3 pieces, you have eaten $\frac{3}{10}$ of the chocolate bar. These real-life examples underscore the importance of fractions in helping us understand and manage parts of a whole in practical situations.

Chapter 10

Equivalent Fractions and Simplification

Definition and Concept of Equivalent Fractions

Equivalent fractions represent the same proportional value even though their numerators and denominators may differ. This property relies on the principle that multiplying or dividing both the numerator and the denominator by the same nonzero number does not change the underlying value of the fraction. For instance, if a fraction is expressed as
$$\frac{a}{b},$$
then the fraction
$$\frac{a \times k}{b \times k} \quad (k \neq 0)$$
is equivalent, since the ratio between numerator and denominator remains constant. In this context, the equality of the fractional values is confirmed when their decimal equivalents or positions on a number line coincide.

Methods for Generating Equivalent Fractions

Equivalent fractions may be generated by scaling the numerator and denominator by a common factor. Multiplying both components of a fraction by the same integer produces a new fraction with larger numbers while preserving the original value. In mathematical terms, the equality

$$\frac{a}{b} = \frac{a \times k}{b \times k}$$

demonstrates this principle. Conversely, when a fraction contains a numerator and a denominator that share common factors, division by these shared factors can be used to reduce the fraction. This process does not alter the fraction's value but merely expresses it in a different, often more convenient, form.

Strategies for Simplifying Fractions

The process of simplification involves reducing a fraction to its simplest form—a state in which the numerator and the denominator have no common factors other than 1. The first step in this process is to determine the greatest common factor (GCF) of the numerator and the denominator. Once the GCF is identified, both the numerator and the denominator are divided by this common number to yield the simplified fraction. For example, consider the fraction

$$\frac{6}{8}.$$

The greatest common factor of 6 and 8 is 2. Dividing both the numerator and the denominator by 2 results in

$$\frac{6 \div 2}{8 \div 2} = \frac{3}{4},$$

which represents the fraction in its simplest form. Such strategies provide an efficient method for reducing fractions to facilitate easier calculation.

Visual Representations and Conceptual Illustrations

Diagrams and visual models are invaluable for illustrating the concept of equivalent fractions and the process of simplification. A graphical representation such as a pie chart, fraction bar, or partitioned rectangle visually demonstrates how a whole can be divided into equal segments. For example, a circle divided into eight congruent parts may be used to depict the fraction

$$\frac{4}{8},$$

with half of the circle shaded. By redrawing the same representation to show that four of eight parts correspond to the same proportion as one of two similar parts, the equivalence of $\frac{4}{8}$ and $\frac{1}{2}$ is clearly exhibited. Such visual techniques provide a concrete connection between abstract numerical operations and real-world divisions.

Worked Examples and Illustrations

Detailed examples help to concretize the process of generating equivalent fractions and simplifying them for ease of calculation. One example considers the fraction

$$\frac{8}{12}.$$

A systematic examination of the factors reveals that both 8 and 12 have a common factor of 4. By dividing the numerator and denominator by 4, the fraction simplifies as follows:

$$\frac{8}{12} = \frac{8 \div 4}{12 \div 4} = \frac{2}{3}.$$

In another example, the fraction

$$\frac{9}{15}$$

is analyzed by noting that both 9 and 15 are divisible by 3. Dividing by 3 yields:

$$\frac{9}{15} = \frac{9 \div 3}{15 \div 3} = \frac{3}{5}.$$

Each example illustrates that although the numerical values in the numerator and denominator change through multiplication or division by a common factor, the fraction retains the same overall value. The systematic approach of identifying common factors and applying a uniform division or multiplication underpins the clarity and efficiency of fraction simplification in various arithmetic operations.

Multiple Choice Questions

1. Which of the following best defines equivalent fractions?

    (a) Fractions that have the same numerator.

    (b) Fractions that have the same denominator.

    (c) Fractions that represent the same proportional value even if their numerators and denominators are different.

    (d) Fractions that can only be simplified by multiplication.

2. To generate an equivalent fraction from a given fraction $\frac{a}{b}$, which operation is performed?

    (a) Multiply only the numerator by a nonzero number.

    (b) Multiply only the denominator by a nonzero number.

    (c) Multiply both the numerator and the denominator by the same nonzero number.

    (d) Add the same nonzero number to both the numerator and denominator.

3. What is the first step in simplifying a fraction to its simplest form?

    (a) Adding 1 to both the numerator and the denominator.

    (b) Finding the greatest common factor (GCF) of the numerator and the denominator.

    (c) Multiplying both the numerator and the denominator by a nonzero number.

    (d) Subtracting the numerator from the denominator.

4. Which of the following is the simplified form of the fraction $\frac{8}{12}$?

(a) $\frac{2}{3}$

(b) $\frac{4}{6}$

(c) $\frac{1}{2}$

(d) $\frac{3}{4}$

5. Which visual model is most effective for understanding equivalent fractions?

    (a) A pie chart divided into equal segments.

    (b) A bar graph representing unrelated data.

    (c) A scatter plot showing data distribution.

    (d) A line graph showing changes over time.

6. To simplify the fraction $\frac{9}{15}$, what is the greatest common factor (GCF) of 9 and 15?

    (a) 5

    (b) 3

    (c) 9

    (d) 15

7. Which property allows a fraction to remain unchanged when its numerator and denominator are both multiplied or divided by the same nonzero number?

    (a) Distributive Property

    (b) Commutative Property

    (c) Multiplicative Scaling Property

    (d) Associative Property

Answers:

1. **C: Fractions that represent the same proportional value even if their numerators and denominators are different.**
Explanation: Even though the numerators and denominators might differ, equivalent fractions represent the same part of a whole. Their values are identical because they express the same ratio.

2. **C: Multiply both the numerator and the denominator by the same nonzero number.**
   Explanation: When both parts of the fraction are multiplied by the same nonzero number, the ratio stays the same; this is the key method for creating an equivalent fraction.

3. **B: Finding the greatest common factor (GCF) of the numerator and the denominator.**
   Explanation: Simplifying a fraction involves identifying the largest number that divides both the numerator and the denominator and then dividing them by that number to reduce the fraction to its simplest form.

4. **A:** $\frac{2}{3}$
   Explanation: The GCF of 8 and 12 is 4. Dividing both the numerator and the denominator by 4 gives $\frac{8 \div 4}{12 \div 4} = \frac{2}{3}$.

5. **A: A pie chart divided into equal segments.**
   Explanation: A pie chart, partitioned into equal parts, visually shows how a whole is subdivided into fractions. This helps students see that different fractions (like $\frac{4}{8}$ and $\frac{1}{2}$) can represent the same portion of the whole.

6. **B: 3**
   Explanation: The common factors of 9 and 15 are 1 and 3, with 3 being the greatest. Dividing both 9 and 15 by 3 simplifies the fraction to $\frac{3}{5}$.

7. **C: Multiplicative Scaling Property**
   Explanation: This property states that if you multiply or divide both the numerator and the denominator by the same nonzero number, the value of the fraction remains unchanged.

Practice Problems

1. Simplify the fraction
$$\frac{6}{8}$$
by identifying the greatest common factor (GCF) of the numerator and denominator and dividing both by it.

2. Write three equivalent fractions for
$$\frac{3}{5}$$
by multiplying both the numerator and the denominator by the same nonzero integer.

3. Explain why the fractions
$$\frac{4}{10} \text{ and } \frac{2}{5}$$
are equivalent. Use the method of simplifying fractions in your explanation.

4. Determine whether the fractions
$$\frac{8}{12} \text{ and } \frac{2}{3}$$
are equivalent by simplifying the first fraction completely.

5. Simplify the fraction
$$\frac{15}{35}$$
completely by finding the greatest common factor of the numerator and denominator.

6. Using a visual representation, describe how the fraction
$$\frac{5}{10}$$
is equivalent to
$$\frac{1}{2}.$$
Include a description of a diagram (such as a pie chart or a number line) that illustrates this equivalence.

Answers

1. **Solution:** To simplify
$$\frac{6}{8},$$
first identify the greatest common factor (GCF) of 6 and 8. The factors of 6 are 1, 2, 3, and 6; the factors of 8 are 1, 2,

4, and 8. The greatest common factor is 2. Divide both the numerator and the denominator by 2:
$$\frac{6 \div 2}{8 \div 2} = \frac{3}{4}.$$
Therefore,
$$\frac{6}{8} = \frac{3}{4}.$$

2. **Solution:** To generate equivalent fractions for
$$\frac{3}{5},$$
multiply both the numerator and the denominator by the same nonzero integer. For example:

(a) Multiply by 2:
$$\frac{3 \times 2}{5 \times 2} = \frac{6}{10}.$$

(b) Multiply by 3:
$$\frac{3 \times 3}{5 \times 3} = \frac{9}{15}.$$

(c) Multiply by 4:
$$\frac{3 \times 4}{5 \times 4} = \frac{12}{20}.$$

Each fraction represents the same value as
$$\frac{3}{5},$$
since multiplying both parts of a fraction by the same number does not change its overall value.

3. **Solution:** To show that
$$\frac{4}{10} \quad \text{and} \quad \frac{2}{5}$$
are equivalent, simplify the fraction
$$\frac{4}{10}.$$
The GCF of 4 and 10 is 2. Dividing the numerator and denominator by 2 gives:
$$\frac{4 \div 2}{10 \div 2} = \frac{2}{5}.$$

Since the simplified form of
$$\frac{4}{10}$$
is
$$\frac{2}{5},$$
the two fractions are equivalent.

4. **Solution:** Consider the fraction
$$\frac{8}{12}.$$
Find the GCF of 8 and 12. The factors of 8 are 1, 2, 4, and 8; for 12 the factors are 1, 2, 3, 4, 6, and 12. The GCF is 4. Dividing both parts by 4 yields:
$$\frac{8 \div 4}{12 \div 4} = \frac{2}{3}.$$
This shows that
$$\frac{8}{12}$$
is equivalent to
$$\frac{2}{3}.$$

5. **Solution:** To simplify
$$\frac{15}{35},$$
first determine the GCF of 15 and 35. The factors of 15 are 1, 3, 5, and 15; the factors of 35 are 1, 5, 7, and 35. The GCF is 5. Dividing both the numerator and the denominator by 5 gives:
$$\frac{15 \div 5}{35 \div 5} = \frac{3}{7}.$$
Therefore, the fraction simplifies to
$$\frac{3}{7}.$$

6. **Solution:** To explain why
$$\frac{5}{10}$$

is equivalent to
$$\frac{1}{2},$$
consider a visual representation such as a pie chart or number line. For a pie chart, imagine a circle divided into 10 equal slices. If 5 of these slices are shaded, then 5 out of 10 parts represent half of the circle because 5 is exactly half of 10. Similarly, on a number line marked from 0 to 10, the point at 5 represents the midpoint, showing that 5/10 equals 1/2. This visual approach confirms that both fractions describe the same portion of a whole.

Chapter 11

Operations with Fractions: Addition and Subtraction

Foundational Concepts in Fraction Operations

Fractions represent parts of a whole, defined by a numerator and a denominator. In operations such as addition and subtraction, it is essential that the fractions describe parts of the same whole. When fractions have different denominators, the sizes of the parts are not equal, and direct arithmetic on the numerators does not yield a meaningful result. The process of aligning the parts of the whole involves converting the fractions to equivalent forms that share a common denominator. This conversion ensures that each fraction represents portions of the whole that are of equal size, thereby allowing reliable addition or subtraction of the numerators.

Identifying and Establishing a Common Denominator

A common denominator is an integral component when performing addition or subtraction with fractions. Two or more fractions require conversion to equivalent forms that all possess the same

denominator. The most efficient approach is to determine the least common multiple (LCM) of the denominators. The LCM provides the smallest number that each original denominator divides into evenly, which simplifies subsequent calculations and helps minimize the need for extra simplification. For example, given fractions such as
$$\frac{1}{3} \text{ and } \frac{1}{4},$$
the denominators 3 and 4 have a least common multiple of 12. Multiplying the numerator and denominator of the first fraction by 4 transforms it into
$$\frac{4}{12},$$
and multiplying the numerator and denominator of the second fraction by 3 transforms it into
$$\frac{3}{12}.$$
The fractions now represent parts of a whole that are all divided into 12 equal pieces.

Addition of Fractions with Common Denominators

When fractions possess a common denominator, the process of addition is straightforward. The technique involves adding the numerators while retaining the common denominator. This method holds true because the denominator specifies the equal parts of the whole, and the numerators specify how many of those parts are being combined. For instance, consider the fractions
$$\frac{3}{8} \text{ and } \frac{2}{8}.$$
Since both fractions share the denominator 8, their sum is computed by adding the numerators:
$$\frac{3}{8} + \frac{2}{8} = \frac{3+2}{8} = \frac{5}{8}.$$
It is important to note that the resulting fraction is in its simplest form when the numerator and denominator no longer share any common factors other than 1.

Subtraction of Fractions with Common Denominators

The process for subtracting fractions that have common denominators follows a similar rationale to addition. With the denominators already aligned, subtraction is performed by subtracting the numerator of the subtrahend from the numerator of the minuend, while the common denominator remains unchanged. For example, in the subtraction
$$\frac{7}{10} - \frac{3}{10},$$
the operation is executed by subtracting 3 from 7 in the numerator:
$$\frac{7}{10} - \frac{3}{10} = \frac{7-3}{10} = \frac{4}{10}.$$
The resulting fraction may then be simplified if the numerator and denominator share any common factors. In this example, both 4 and 10 are divisible by 2, and simplifying the fraction yields
$$\frac{4 \div 2}{10 \div 2} = \frac{2}{5}.$$

Worked Examples in Fraction Addition and Subtraction

Thorough examples illustrate the procedures for converting fractions to equivalent forms with common denominators and then performing addition or subtraction.

Example 1: Adding Fractions with Different Denominators

Consider the fractions
$$\frac{1}{3} \text{ and } \frac{1}{4}.$$
The least common denominator is determined as 12. The first fraction is converted by multiplying both numerator and denominator by 4:
$$\frac{1 \times 4}{3 \times 4} = \frac{4}{12},$$

and the second fraction is converted by multiplying both numerator and denominator by 3:

$$\frac{1 \times 3}{4 \times 3} = \frac{3}{12}.$$

Both fractions now have a denominator of 12, and their sum is given by:

$$\frac{4}{12} + \frac{3}{12} = \frac{4+3}{12} = \frac{7}{12}.$$

Example 2: Subtracting Fractions with a Common Denominator

Examine the fractions

$$\frac{9}{15} \text{ and } \frac{4}{15}.$$

With a shared denominator of 15, subtraction is performed directly:

$$\frac{9}{15} - \frac{4}{15} = \frac{9-4}{15} = \frac{5}{15}.$$

The fraction can be simplified by noting that both 5 and 15 have a common factor of 5, resulting in:

$$\frac{5 \div 5}{15 \div 5} = \frac{1}{3}.$$

Example 3: Adding Fractions with a Simplification Step

Consider the fractions

$$\frac{2}{5} \text{ and } \frac{3}{7}.$$

The common denominator is found by computing the product, which is 35. The first fraction is converted by multiplying numerator and denominator by 7:

$$\frac{2 \times 7}{5 \times 7} = \frac{14}{35},$$

and the second fraction is converted by multiplying numerator and denominator by 5:

$$\frac{3 \times 5}{7 \times 5} = \frac{15}{35}.$$

With the common denominator in place, addition yields:
$$\frac{14}{35} + \frac{15}{35} = \frac{14+15}{35} = \frac{29}{35}.$$

Since 29 and 35 do not have a common factor other than 1, the fraction is already in its simplest form.

Each method illustrated reinforces the critical idea that equivalent fractions must be established through a common denominator before engaging in addition or subtraction. The systematic approach of aligning denominators ensures consistency and accuracy in the calculation process, forming a reliable foundation for further arithmetic operations with fractions.

Multiple Choice Questions

1. When adding or subtracting fractions, why must the fractions have a common denominator?

    (a) Because it changes the value of each fraction.

    (b) Because it aligns the parts of the whole so they are equal in size.

    (c) Because it makes the numerators equal.

    (d) Because it converts the fractions into decimals.

2. What is the most efficient method for finding a common denominator for fractions with different denominators?

    (a) Adding the denominators.

    (b) Multiplying the numerators.

    (c) Finding the least common multiple (LCM) of the denominators.

    (d) Dividing the larger denominator by the smaller one.

3. When adding fractions that already have the same denominator, what is the correct procedure?

    (a) Add the numerators and add the denominators.

    (b) Multiply the numerators and keep the denominator unchanged.

    (c) Add the numerators and retain the common denominator.

(d) Subtract the numerators and keep the denominator unchanged.

4. What is the sum of the fractions
$$\frac{3}{8} + \frac{2}{8}?$$

   (a) $\frac{5}{8}$
   (b) $\frac{6}{16}$
   (c) $\frac{3}{10}$
   (d) $\frac{5}{10}$

5. To convert the fractions
$$\frac{1}{3} \text{ and } \frac{1}{4}$$
   to equivalent fractions with a common denominator, which conversion is correct?

   (a) $\frac{1}{3} = \frac{4}{12}$ and $\frac{1}{4} = \frac{3}{12}$
   (b) $\frac{1}{3} = \frac{3}{9}$ and $\frac{1}{4} = \frac{4}{9}$
   (c) $\frac{1}{3} = \frac{2}{6}$ and $\frac{1}{4} = \frac{2}{6}$
   (d) $\frac{1}{3} = \frac{3}{12}$ and $\frac{1}{4} = \frac{4}{12}$

6. Which describes the proper method for subtracting fractions that have a common denominator?

   (a) Subtract the numerators and keep the denominator unchanged.
   (b) Subtract the denominators and keep the numerator unchanged.
   (c) Add the numerators, then subtract the denominator.
   (d) Multiply the numerators and divide by the denominator.

7. After subtracting fractions with a common denominator, you obtain the result
$$\frac{4}{10}.$$
   What is its simplest form?

   (a) $\frac{2}{5}$

(b) $\frac{4}{10}$

(c) $\frac{1}{2}$

(d) $\frac{3}{5}$

Answers:

1. **B: Because it aligns the parts of the whole so they are equal in size** Explanation: When fractions have the same denominator, the pieces of the whole they represent are of equal size. This alignment is essential for correctly adding or subtracting the numerators.

2. **C: Finding the least common multiple (LCM) of the denominators** Explanation: The LCM of the denominators gives the smallest common denominator that both fractions can be converted to, making calculations simpler and reducing the need for further simplification.

3. **C: Add the numerators and retain the common denominator** Explanation: With a common denominator, you add only the numerators while keeping the denominator the same because the denominators already represent identical partitions of the whole.

4. **A:** $\frac{5}{8}$ Explanation: Since the denominators are already the same, simply add the numerators (3 + 2) to obtain 5, with the denominator remaining 8.

5. **A:** $\frac{1}{3} = \frac{4}{12}$ **and** $\frac{1}{4} = \frac{3}{12}$ Explanation: Multiplying the numerator and denominator of $\frac{1}{3}$ by 4, and of $\frac{1}{4}$ by 3, gives the equivalent fractions with a common denominator of 12, which is the LCM of 3 and 4.

6. **A: Subtract the numerators and keep the denominator unchanged** Explanation: When subtracting fractions with the same denominator, subtract the numerator of the second fraction from the numerator of the first, while the common denominator remains unchanged.

7. **A:** $\frac{2}{5}$ Explanation: The fraction $\frac{4}{10}$ can be simplified by dividing both the numerator and the denominator by their greatest common divisor, which is 2. Thus, $\frac{4}{10}$ simplifies to $\frac{2}{5}$.

Practice Problems

1. Add the fractions:
$$\frac{3}{8} + \frac{2}{8}$$

2. Subtract the fractions:
$$\frac{9}{15} - \frac{4}{15}$$

3. Add the fractions with different denominators:
$$\frac{1}{3} + \frac{1}{4}$$

4. Subtract the fractions with different denominators:
$$\frac{5}{6} - \frac{1}{4}$$

5. Compute the sum of the fractions and simplify if necessary:
$$\frac{2}{5} + \frac{3}{7}$$

6. Evaluate the expression:
$$\frac{7}{12} + \frac{1}{4} - \frac{5}{8}$$

Answers

1. **Solution:** Since the fractions have the same denominator, we add the numerators directly. Thus,
$$\frac{3}{8} + \frac{2}{8} = \frac{3+2}{8} = \frac{5}{8}.$$
The final answer is $\frac{5}{8}$.

2. **Solution:** With a common denominator, subtraction is done by subtracting the numerators. We have:
$$\frac{9}{15} - \frac{4}{15} = \frac{9-4}{15} = \frac{5}{15}.$$
Next, simplify by dividing the numerator and denominator by their greatest common factor, which is 5:
$$\frac{5 \div 5}{15 \div 5} = \frac{1}{3}.$$
Therefore, the final answer is $\frac{1}{3}$.

3. **Solution:** When adding fractions with different denominators, the first step is to find a common denominator. The least common multiple (LCM) of 3 and 4 is 12. Convert each fraction:
$$\frac{1}{3} = \frac{1 \times 4}{3 \times 4} = \frac{4}{12} \text{ and } \frac{1}{4} = \frac{1 \times 3}{4 \times 3} = \frac{3}{12}.$$
Now add the fractions:
$$\frac{4}{12} + \frac{3}{12} = \frac{4+3}{12} = \frac{7}{12}.$$
The final answer is $\frac{7}{12}$.

4. **Solution:** For subtraction with different denominators, find the least common denominator. The LCM of 6 and 4 is 12. Convert each fraction accordingly:
$$\frac{5}{6} = \frac{5 \times 2}{6 \times 2} = \frac{10}{12} \text{ and } \frac{1}{4} = \frac{1 \times 3}{4 \times 3} = \frac{3}{12}.$$
Subtract the fractions:
$$\frac{10}{12} - \frac{3}{12} = \frac{10-3}{12} = \frac{7}{12}.$$
The final answer is $\frac{7}{12}$.

5. **Solution:** Here the denominators are 5 and 7. Their least common multiple is 35. Convert each fraction:

$$\frac{2}{5} = \frac{2 \times 7}{5 \times 7} = \frac{14}{35} \quad \text{and} \quad \frac{3}{7} = \frac{3 \times 5}{7 \times 5} = \frac{15}{35}.$$

Add the fractions:

$$\frac{14}{35} + \frac{15}{35} = \frac{14 + 15}{35} = \frac{29}{35}.$$

Since 29 and 35 have no common factors other than 1, the fraction is in simplest form. The final answer is $\frac{29}{35}$.

6. **Solution:** In this mixed operation, we first determine the least common denominator for the fractions with denominators 12, 4, and 8. The least common denominator is 24. Convert each fraction:

$$\frac{7}{12} = \frac{7 \times 2}{12 \times 2} = \frac{14}{24},$$

$$\frac{1}{4} = \frac{1 \times 6}{4 \times 6} = \frac{6}{24},$$

$$\frac{5}{8} = \frac{5 \times 3}{8 \times 3} = \frac{15}{24}.$$

Next, perform the operations in order:

$$\frac{14}{24} + \frac{6}{24} = \frac{20}{24},$$

then subtract:

$$\frac{20}{24} - \frac{15}{24} = \frac{5}{24}.$$

The final answer is $\frac{5}{24}$.

Chapter 12

Operations with Fractions: Multiplication and Division

Multiplication of Fractions

1 Multiplying Numerators and Denominators

Multiplication of fractions is performed by directly multiplying the numerators to form the numerator of the product and multiplying the denominators to form the denominator of the product. For instance, the multiplication

$$\frac{2}{3} \times \frac{4}{5}$$

involves calculating

$$\frac{2 \times 4}{3 \times 5} = \frac{8}{15}.$$

This method does not require any adjustments to the denominators, as is necessary with addition and subtraction, and it clearly illustrates the inherent multiplicative properties of fractions.

2 Simplification through Cancellation of Common Factors

Before performing the multiplication, simplification through cancellation of common factors between a numerator of one fraction and a denominator of the other can be applied. This process, often called cross-cancellation, minimizes the need for further reduction after the multiplication. Consider the example

$$\frac{3}{8} \times \frac{4}{9}.$$

In this multiplication, the numerator 3 and the denominator 9 share a common factor of 3. Dividing these by 3 gives

$$\frac{3 \div 3}{9 \div 3} = \frac{1}{3}.$$

Similarly, the numerator 4 and the denominator 8 share a common factor of 4. Dividing these by 4 results in

$$\frac{4 \div 4}{8 \div 4} = \frac{1}{2}.$$

Thus, the multiplication simplifies to

$$\frac{1}{2} \times \frac{1}{3} = \frac{1 \times 1}{2 \times 3} = \frac{1}{6}.$$

This cancellation technique simplifies the computation and produces the final product in its simplest form.

Division of Fractions

1 Reciprocals in the Division Process

Division of fractions is accomplished by utilizing the reciprocal of the divisor. The reciprocal of a fraction is obtained by interchanging its numerator and denominator. For any fraction

$$\frac{a}{b},$$

with nonzero values for a and b, its reciprocal is

$$\frac{b}{a}.$$

This concept transforms the operation of division into a multiplication problem. Dividing by a fraction becomes equivalent to multiplying by its reciprocal, which simplifies the division process markedly.

2 Performing Division by Multiplying by the Reciprocal

To divide one fraction by another, the procedure begins by replacing the division sign with a multiplication sign and then substituting the divisor with its reciprocal. For example, the division

$$\frac{5}{6} \div \frac{2}{3}$$

is rewritten as

$$\frac{5}{6} \times \frac{3}{2}.$$

Multiplication of the fractions proceeds as follows:

$$\frac{5 \times 3}{6 \times 2} = \frac{15}{12}.$$

The resulting fraction is simplified by dividing both the numerator and the denominator by their common factor, in this case 3:

$$\frac{15 \div 3}{12 \div 3} = \frac{5}{4}.$$

This procedure, which relies on the recognition and application of the reciprocal, provides a systematic approach to the division of fractions and reinforces the connection between division and multiplication.

Multiple Choice Questions

1. When multiplying the fractions $\frac{2}{3}$ and $\frac{4}{5}$, which operation gives the correct product?

    (a) $\frac{2+4}{3+5}$
    (b) $\frac{2+4}{3\times 5}$
    (c) $\frac{2\times 4}{3\times 5}$
    (d) $\frac{2\times 5}{3\times 4}$

2. When multiplying $\frac{3}{8}$ by $\frac{4}{9}$, which cancellation is correctly applied before multiplying?

   (a) Cancel the 3 in the numerator of the first fraction with the 8 in its denominator.

   (b) Cancel the 3 in the numerator of the first fraction with the 9 in the denominator of the second, and cancel the 4 in the numerator of the second with the 8 in the denominator of the first.

   (c) Cancel the 4 in the numerator of the second fraction with the 9 in its denominator.

   (d) No cancellation is possible because the fractions are already in simplest form.

3. Cross-cancellation in fraction multiplication is best described as:

   (a) Multiplying the numerators across and the denominators across.

   (b) Dividing a numerator of one fraction and a denominator of the other by their common factor before multiplying.

   (c) Adding a common factor to both fractions to simplify multiplication.

   (d) Replacing each fraction with its reciprocal before multiplying.

4. To compute the division $\frac{5}{6} \div \frac{2}{3}$, what is the correct process?

   (a) Multiply $\frac{5}{6}$ by $\frac{2}{3}$.

   (b) Multiply $\frac{5}{6}$ by the reciprocal of $\frac{2}{3}$.

   (c) Subtract $\frac{2}{3}$ from $\frac{5}{6}$.

   (d) Invert both fractions and then multiply.

5. After rewriting $\frac{5}{6} \div \frac{2}{3}$ as a multiplication problem, which fraction replaces $\frac{2}{3}$?

   (a) $\frac{2}{3}$

   (b) $\frac{3}{2}$

   (c) $\frac{6}{5}$

   (d) $\frac{5}{6}$

6. What is the reciprocal of the fraction $\frac{7}{9}$?

    (a) $\frac{7}{9}$
    (b) $\frac{9}{7}$
    (c) $\frac{1}{7}$
    (d) $\frac{1}{9}$

7. After simplifying by cancellation, multiplying $\frac{1}{2}$ by $\frac{1}{3}$ results in:

    (a) $\frac{1}{5}$
    (b) $\frac{1}{6}$
    (c) $\frac{1}{8}$
    (d) $\frac{2}{3}$

Answers:

1. **C:** $\frac{2\times 4}{3\times 5}$
   Multiplying fractions requires you to multiply the numerators together and the denominators together: $2\times 4 = 8$ and $3\times 5 = 15$, giving $\frac{8}{15}$.

2. **B: Cancel the 3 with the 9 and the 4 with the 8**
   In $\frac{3}{8} \times \frac{4}{9}$, notice that 3 and 9 share a common factor of 3, and 4 and 8 share a common factor of 4. Cancelling these before multiplying simplifies the product to $\frac{1}{2} \times \frac{1}{3} = \frac{1}{6}$.

3. **B: Dividing a numerator of one fraction and a denominator of the other by their common factor before multiplying**
   Cross-cancellation speeds up multiplication by reducing each fraction to simpler forms, making the arithmetic easier.

4. **B: Multiply $\frac{5}{6}$ by the reciprocal of $\frac{2}{3}$**
   Dividing by a fraction is equivalent to multiplying by its reciprocal. The reciprocal of $\frac{2}{3}$ is $\frac{3}{2}$, so the operation becomes $\frac{5}{6} \times \frac{3}{2}$.

5. **B:** $\frac{3}{2}$
   The reciprocal of a fraction is formed by interchanging the numerator and denominator. Thus, the reciprocal of $\frac{2}{3}$ is $\frac{3}{2}$.

6. **B:** $\frac{9}{7}$

   The reciprocal of $\frac{7}{9}$ is obtained by swapping the numerator and denominator, which gives $\frac{9}{7}$.

7. **B:** $\frac{1}{6}$

   Multiplying the simplified fractions $\frac{1}{2}$ and $\frac{1}{3}$ results in $\frac{1\times 1}{2\times 3} = \frac{1}{6}$.

Practice Problems

1. Multiply the following fractions and simplify your answer:

$$\frac{7}{9} \times \frac{3}{14}$$

2. Multiply the following fractions:

$$\frac{2}{5} \times \frac{7}{3}$$

3. Solve the division problem:

$$\frac{5}{8} \div \frac{3}{4}$$

4. Solve the division problem:
$$\frac{3}{7} \div \frac{9}{14}$$

5. Explain why dividing by a fraction is equivalent to multiplying by its reciprocal. Provide a detailed explanation in your answer.

6. Evaluate the following expression that involves both division

and multiplication:
$$\left(\frac{3}{4} \div \frac{2}{5}\right) \times \frac{7}{8}$$

Answers

1. **Question:** Multiply the following fractions:
$$\frac{7}{9} \times \frac{3}{14}$$

   Solution: To multiply these fractions, we can simplify first by using cross-cancellation. Notice that 7 (from the first fraction) and 14 (from the second fraction) share a common factor of 7. Dividing, we have:
$$\frac{7 \div 7}{14 \div 7} = \frac{1}{2}.$$

   Next, 3 (from the second fraction) and 9 (from the first fraction) share a common factor of 3:
$$\frac{3 \div 3}{9 \div 3} = \frac{1}{3}.$$

   The multiplication then becomes:
$$\frac{1}{3} \times \frac{1}{2} = \frac{1 \times 1}{3 \times 2} = \frac{1}{6}.$$

   Therefore, the simplified result is:
$$\frac{1}{6}.$$

2. **Question:** Multiply the following fractions:
$$\frac{2}{5} \times \frac{7}{3}$$

Solution: Multiply the numerators together and the denominators together:
$$\frac{2 \times 7}{5 \times 3} = \frac{14}{15}.$$

No cancellation is possible between any numerator and denominator in this multiplication, so the fraction is already in simplest form. Thus, the product is:
$$\frac{14}{15}.$$

3. **Question:** Solve the division problem:
$$\frac{5}{8} \div \frac{3}{4}$$

Solution: Dividing by a fraction is equivalent to multiplying by its reciprocal. Replace the division with multiplication using the reciprocal of $\frac{3}{4}$:
$$\frac{5}{8} \times \frac{4}{3}.$$

Multiply the fractions:
$$\frac{5 \times 4}{8 \times 3} = \frac{20}{24}.$$

Simplify by dividing numerator and denominator by 4:
$$\frac{20 \div 4}{24 \div 4} = \frac{5}{6}.$$

Therefore, the answer is:
$$\frac{5}{6}.$$

4. **Question:** Solve the division problem:
$$\frac{3}{7} \div \frac{9}{14}$$

Solution: Begin by rewriting the division as multiplication by the reciprocal of the divisor:
$$\frac{3}{7} \times \frac{14}{9}.$$
Notice that 14 and 7 have a common factor; dividing 14 by 7 gives 2:
$$\frac{3}{1} \times \frac{2}{9} = \frac{6}{9}.$$
Then simplify $\frac{6}{9}$ by dividing both numerator and denominator by 3:
$$\frac{6 \div 3}{9 \div 3} = \frac{2}{3}.$$
Thus, the simplified result is:
$$\frac{2}{3}.$$

5. **Question:** Explain why dividing by a fraction is equivalent to multiplying by its reciprocal. **Solution:** When we divide by a fraction, we are essentially determining how many times that fraction fits into another number. Mathematics defines division as the inverse operation of multiplication. For any nonzero fraction
$$\frac{a}{b},$$
its reciprocal is given by
$$\frac{b}{a}.$$
This is because:
$$\frac{a}{b} \times \frac{b}{a} = 1.$$
Therefore, dividing by $\frac{a}{b}$ is the same as multiplying by $\frac{b}{a}$:
$$\frac{c}{d} \div \frac{a}{b} = \frac{c}{d} \times \frac{b}{a}.$$
This method converts a division problem into an easier and more straightforward multiplication problem, reducing the complexity often encountered with division of fractions.

6. **Question:** Evaluate the following expression:
$$\left(\frac{3}{4} \div \frac{2}{5}\right) \times \frac{7}{8}$$

Solution: Begin by solving the division inside the parentheses. Replace the division with multiplication by the reciprocal:
$$\frac{3}{4} \div \frac{2}{5} = \frac{3}{4} \times \frac{5}{2}.$$

Multiply the fractions:
$$\frac{3 \times 5}{4 \times 2} = \frac{15}{8}.$$

Next, multiply this result by $\frac{7}{8}$:
$$\frac{15}{8} \times \frac{7}{8} = \frac{15 \times 7}{8 \times 8} = \frac{105}{64}.$$

The fraction $\frac{105}{64}$ is already in simplest form. Therefore, the final answer is:
$$\frac{105}{64}.$$

Chapter 13

Introduction to Decimals and Place Value

Transitioning from Fractions to Decimals

Fractions and decimals represent two interrelated methods of expressing parts of a whole. Fractions denote the ratio between two integers, whereas decimals express the same quantity using the base-ten numeral system. Converting a fraction to its decimal form involves dividing the numerator by the denominator. For example, the fraction 1/2 becomes 0.5 after the division process, and the fraction 3/4 converts to 0.75. This transformation from fractions to decimals not only provides an alternative numerical expression but also aligns with a consistent method for representing numbers in a base-ten context. The division process may yield terminating decimals when the quotient resolves after a certain number of digits or repeating decimals when a cyclic pattern appears. Such conversions highlight the intrinsic relationship between fractional representations and the standardized base-ten system.

Understanding Decimal Notation

Decimal notation is a method of writing numbers that encompasses both whole numbers and fractional parts separated by a decimal point. Each digit in a decimal number possesses a value determined by its position relative to the decimal point. Digits to the left of the decimal point represent whole number values in powers of ten, while digits to the right indicate parts of a whole, corresponding to 10 raised to negative exponents. For instance, in the decimal number 3.47, the digit 3 occupies the ones place, the digit 4 is in the tenths place (representing 4/10), and the digit 7 is in the hundredths place (representing 7/100). The clear separation into an integral part and a fractional part facilitates the expression of any number in a unified format. This arrangement underlies operations in arithmetic by ensuring that both whole numbers and fractional quantities are managed coherently within the same computational framework.

Place Value in a Base-Ten System

In a base-ten, or decimal, system, the value of each digit depends on its specific location within the number. Each digit is multiplied by a power of ten corresponding to its place value. Consider the number 452.67: the digit 4 is in the hundreds place, meaning its value is 4×10^2 (or 400); the digit 5 is in the tens place, contributing 5×10^1 (or 50); and the digit 2 occupies the ones place, valued at 2×10 (or 2). Following the decimal point, the digit 6 is in the tenths place, equal to 6×10^1 (or 0.6), and the digit 7 is in the hundredths place, equivalent to 7×10^2 (or 0.07). This system of place value allows any decimal number to be expressed as the sum of its individual components, where the positional weighting of each digit plays a critical role. Moreover, the placement of digits to the right of the decimal point directly suggests the formation of fractions with denominators that are powers of ten; for example, 0.67 can be represented as 67/100. Such an understanding of place value forms the cornerstone for working with decimals and reinforces the structural logic inherent in the base-ten numerical system.

Multiple Choice Questions

1. What is the process used to convert a fraction to a decimal?

    (a) Multiplying the numerator by the denominator

    (b) Dividing the numerator by the denominator

- (c) Adding the numerator and the denominator
- (d) Subtracting the denominator from the numerator

2. Which of the following fractions is most likely to result in a repeating decimal when converted?

    - (a) 1/2
    - (b) 3/4
    - (c) 1/3
    - (d) 5/10

3. In the decimal number 3.47, what does the digit "4" represent?

    - (a) Ones place
    - (b) Tenths place
    - (c) Hundreds place
    - (d) Hundredths place

4. In a base-ten system, which power of ten corresponds to the hundredths place?

    - (a) 10^1
    - (b) 10^0
    - (c) 10^{-1}
    - (d) 10^{-2}

5. Which statement best describes the significance of place value in a decimal number?

    - (a) Each digit has a fixed value regardless of its position.
    - (b) The value of each digit is determined by its position relative to the decimal point.
    - (c) Every digit contributes equally to the overall value of the number.
    - (d) Only the digits to the left of the decimal point determine the number's value.

6. How can the decimal 0.67 be expressed as a fraction?

    - (a) 67/10

(b) 67/100

   (c) 67/1000

   (d) 6/7

7. Why is understanding decimal notation important in arithmetic operations?

    (a) It simplifies the process of solving complex equations.

    (b) It helps separate whole numbers from fractional parts, ensuring accurate calculations.

    (c) It eliminates the need for converting numbers.

    (d) It only applies to the multiplication and division operations.

Answers:

1. **B: Dividing the numerator by the denominator**
   Converting a fraction to a decimal involves dividing the numerator by the denominator, which reexpresses the fraction in base-ten form.

2. **C: 1/3**
   The fraction 1/3 yields a repeating decimal because the division produces a never-ending cycle of digits.

3. **B: Tenths place**
   In 3.47, the digit "4" is immediately after the decimal point, which represents the tenths place (or 4/10).

4. **D: 10^{-2}**
   The hundredths place corresponds to 10^{-2} because each digit after the decimal point denotes a power of ten with a negative exponent.

5. **B: The value of each digit is determined by its position relative to the decimal point**
   Place value explains that the value of a digit depends on where it is located in the number—digits to the left of the decimal represent whole numbers, while those to the right represent fractional parts.

6. **B: 67/100**
   Since 0.67 has two digits after the decimal point, it represents 67 hundredths, which is written as the fraction 67/100.

7. **B: It helps separate whole numbers from fractional parts, ensuring accurate calculations**
Understanding decimal notation is crucial because it clarifies the roles of various digits (based on their positions), which in turn supports precise arithmetic operations.

Practice Problems

1. Convert the fraction:
$$\frac{3}{8}$$
into its decimal form. Determine whether the resulting decimal is terminating or repeating.

2. Write the decimal number:
$$7.204$$
in expanded form, clearly showing the place value of each digit.

3. Convert the decimal:
$$0.375$$

into a fraction in its simplest form.

4. In the decimal number:
$$82.46$$
identify the place value of the digit 4 and explain what this place value represents.

5. Divide the fraction:
$$\frac{5}{6}$$
by the fraction:
$$\frac{2}{3}$$
then convert the result into a decimal.

6. Convert the repeating decimal:

$$0.727272\ldots$$

into a fraction, providing a step-by-step explanation of the process.

Answers

1. **Solution:**
   To convert the fraction
   $$\frac{3}{8}$$
   to a decimal, divide 3 by 8 using long division:
   $$3 \div 8 = 0.375.$$
   The division terminates because the remainder becomes 0. Therefore, the decimal is terminating.

2. **Solution:**
   The decimal
   $$7.204$$
   can be written in expanded form by expressing each digit according to its place value:
   $$7.204 = 7 + \frac{2}{10} + \frac{0}{100} + \frac{4}{1000}.$$
   Here, 7 represents the ones, 2 is in the tenths place, 0 in the hundredths place, and 4 in the thousandths place.

3. **Solution:**
   The decimal
   $$0.375$$

has three digits after the decimal point, so it can be written as:
$$0.375 = \frac{375}{1000}.$$
Simplify the fraction by dividing the numerator and denominator by 125:
$$\frac{375 \div 125}{1000 \div 125} = \frac{3}{8}.$$
Thus,
$$0.375 = \frac{3}{8}.$$

4. **Solution:**
   In the decimal number
   $$82.46,$$
   the digit 4 is located immediately after the decimal point, which places it in the tenths position. Its value is calculated as:
   $$4 \times \frac{1}{10} = 0.4.$$
   This shows that the digit 4 represents four-tenths of a unit.

5. **Solution:**
   To divide
   $$\frac{5}{6}$$
   by
   $$\frac{2}{3},$$
   multiply the first fraction by the reciprocal of the second:
   $$\frac{5}{6} \div \frac{2}{3} = \frac{5}{6} \times \frac{3}{2} = \frac{15}{12}.$$
   Simplify the fraction by dividing numerator and denominator by 3:
   $$\frac{15 \div 3}{12 \div 3} = \frac{5}{4}.$$
   Converting
   $$\frac{5}{4}$$
   to a decimal:
   $$\frac{5}{4} = 1.25.$$

6. **Solution:**
    Let
    $$x = 0.727272\ldots.$$

    Since the repeating block "72" contains 2 digits, multiply both sides by 100:
    $$100x = 72.727272\ldots.$$

    Subtract the original equation from this new equation:
    $$100x - x = 72.727272\ldots - 0.727272\ldots,$$

    which simplifies to:
    $$99x = 72.$$

    Solve for x:
    $$x = \frac{72}{99}.$$

    Simplify the fraction by dividing numerator and denominator by 9:
    $$\frac{72 \div 9}{99 \div 9} = \frac{8}{11}.$$

    Therefore,
    $$0.727272\ldots = \frac{8}{11}.$$

Chapter 14

Comparing, Rounding, and Ordering Decimals

Comparing Decimals

Comparing decimal numbers requires careful attention to the place value of each digit. The process begins by aligning the decimal points so that the whole number and fractional parts of each number line up exactly. When decimals are written in this manner, the comparison is made by starting at the leftmost digit and moving to the right one position at a time.

For example, when evaluating two decimals such as 3.45 and 3.410, it is helpful to rewrite 3.45 as 3.450 so that both numbers display the same number of digits after the decimal point. The comparison then starts with the ones place; if these are equal, the next step is to compare the tenths place, and this continues sequentially. In the case of 3.450 versus 3.410, after confirming that the digits in the ones and tenths positions are identical, the hundredths place is examined. The digit 5 in the first decimal exceeds the digit 1 in the second decimal, thereby establishing that 3.450 is greater than 3.410. This method of sequential digit comparison ensures that the inherent value of each decimal is assessed in a logical and systematic manner.

In addition, the procedure includes normalizing the numbers by appending zeros when necessary. This practice guarantees that each decimal number is evaluated with the same level of precision, eliminating any confusion that may arise from differing numbers of digits. The intrinsic nature of the decimal system, in which each position represents a power of ten, provides a definitive framework for comparing the relative magnitudes of different decimal numbers.

Rounding Decimals

Rounding decimals is a technique used to simplify a number while retaining a value that is close to the original. The method involves reducing the number of digits to a specified degree of precision, thereby obtaining an approximation that is easier to work with. The process centers on identifying the target place value, and then examining the digit immediately to its right.

Consider a decimal such as 4.367 that is to be rounded to the nearest tenth. The digit in the tenths place is identified, and the digit in the hundredths place is used to determine whether to increase the tenths digit. In this example, the hundredths digit is 6; since this digit is above the threshold value of 5, the tenths digit is raised by one unit. The result of rounding 4.367 to the nearest tenth is therefore 4.4. In a different scenario, a decimal such as 7.142 might be rounded to the nearest hundredth. Here, the digit in the thousandths position, which is 2, is examined. As 2 is below 5, the hundredths digit remains unchanged, and the rounded value becomes 7.14.

Rounding decimals involves a deliberate step-by-step evaluation of the digits, ensuring that the final approximate value accurately reflects the original number within the desired precision. This systematic process highlights the importance of place value and the rules that govern rounding, rendering the concept both practical and essential for a variety of mathematical computations.

Ordering Decimals

Ordering decimals entails arranging a set of numbers in either increasing or decreasing order. The process is streamlined by first ensuring that every number in the set is expressed with the same number of digits after the decimal point. This normalization is achieved by appending zeros to any number that has fewer digits, so that the comparison between digits is performed consistently from the highest place value to the lowest.

For instance, consider the decimals 2.5, 2.45, and 2.405. Rewriting these numbers with three digits after the decimal point results in 2.500, 2.450, and 2.405. With this common format established, the procedure begins with comparing the whole number portion. If these are identical, the analysis continues to the tenths place, and subsequently to the hundredths and thousandths places. In this example, 2.405 is identified as the smallest number because its digits in the tenths and subsequent places are less than those in the other numbers. Following this careful examination, the arranged sequence from least to greatest is determined to be 2.405, 2.450, and 2.500.

The method of standardizing decimal representations before ordering them relies on the structural framework of the base-ten system. By treating missing digits as zeros, the comparison accounts for differences in numerical precision accurately. Consequently, this technique provides a reliable approach for organizing decimal numbers in a clear and logical sequence.

Multiple Choice Questions

1. When comparing decimals, why do we sometimes rewrite a number such as 3.45 as 3.450?

    (a) To make the number larger.
    (b) To ensure that both numbers have the same number of digits for accurate place-value comparison.
    (c) To round the number to the nearest whole number.
    (d) To change the value of the number.

2. When comparing two decimals like 3.450 and 3.410, what is the correct procedure?

    (a) Compare starting from the leftmost digit, ensuring the decimal points are aligned and all decimals have the same number of digits.
    (b) Round both numbers to the nearest tenth, then compare.
    (c) Remove the fractional parts and compare only the whole number parts.
    (d) Multiply both numbers by 10 to eliminate the decimal point and then compare.

3. When rounding the decimal 4.367 to the nearest tenth, which digit determines whether the tenths place should be increased?

(a) The ones digit.

(b) The tenths digit.

(c) The hundredths digit.

(d) The thousandths digit.

4. What is the rounded value of 7.142 when rounding to the nearest hundredth?

    (a) 7.14

    (b) 7.15

    (c) 7.142

    (d) 7.1

5. When ordering the decimals 2.5, 2.45, and 2.405 from least to greatest, what is a critical first step?

    (a) Rounding all numbers to the nearest whole number.

    (b) Removing the decimal points.

    (c) Normalizing the decimals by rewriting them with the same number of digits after the decimal (appending zeros if needed).

    (d) Converting the decimals into fractions.

6. After normalizing the decimals 2.5, 2.45, and 2.405 as 2.500, 2.450, and 2.405 respectively, which is the smallest value?

    (a) 2.5

    (b) 2.45

    (c) 2.405

    (d) All have the same value.

7. Which of the following statements is TRUE regarding the methods for comparing, rounding, and ordering decimals?

    (a) When rounding decimals, the digit to the left of the target place value is used for the rounding decision.

    (b) In ordering decimals, only the whole number parts need to be compared.

    (c) Normalizing decimals by appending zeros ensures a fair, digit-by-digit comparison.

(d) When comparing decimals, it is acceptable to ignore the digits after the decimal point.

Answers:

1. **B:** Appending zeros so that both decimals have the same number of digits allows for an accurate comparison of corresponding place values.

2. **A:** Aligning the decimal points and ensuring equal precision (by adding zeros, if necessary) lets you compare each digit sequentially from left to right.

3. **C:** To round to the nearest tenth, you examine the hundredths digit. In 4.367, the hundredths digit is 6 (which is 5 or greater), so the tenths digit is increased by 1.

4. **A:** When rounding 7.142 to the nearest hundredth, look at the thousandths digit (2). Since 2 is less than 5, the hundredths digit remains unchanged, resulting in 7.14.

5. **C:** Normalizing the decimals by rewriting them with the same number of digits after the decimal point (e.g., writing 2.5 as 2.500) allows you to accurately compare each digit position.

6. **C:** Once normalized, 2.5 becomes 2.500, 2.45 becomes 2.450, and 2.405 remains 2.405. Comparing place values shows that 2.405 is the smallest since its digits in the tenths and subsequent positions are lower.

7. **C:** By appending zeros, each decimal is expressed to the same level of precision. This ensures that the place-value comparison is accurate and fair when comparing, rounding, or ordering decimals.

Practice Problems

1. Compare the decimals

$$3.45 \quad \text{and} \quad 3.410$$

After normalizing these decimals, determine which value is greater.

2. Round the number
$$4.367$$
to the nearest tenth.

3. Round the number
$$7.142$$
to the nearest hundredth.

4. Order the following set of decimals from least to greatest:
$$2.5, \quad 2.45, \quad 2.405$$
Make sure to normalize them by adding zeros where necessary.

5. Compare the decimals
$$3.4, \quad 3.40, \quad 3.400$$
and explain whether they are equivalent after normalization.

6. A number is rounded to the nearest hundredth. Round
$$5.6789$$
and explain the process used to obtain the correct value.

Answers

1. **Solution:**
   To compare
   $$3.45 \quad \text{and} \quad 3.410,$$
   first rewrite 3.45 as 3.450 so that both numbers have the same number of digits after the decimal point. Then compare digit by digit: the ones digits are equal (both 3), the tenths digits are equal (both 4), and the hundredths digits are 5 and 1 respectively. Since 5 is greater than 1, we conclude that
   $$3.450 > 3.410.$$

Therefore, 3.45 is greater than 3.410.

2. **Solution:**
To round 4.367 to the nearest tenth, identify the tenths digit (which is 3) and then check the hundredths digit (which is 6). Since 6 is greater than or equal to 5, increase the tenths digit by one. Thus, 4.367 rounded to the nearest tenth is

$$4.4.$$

3. **Solution:**
For rounding 7.142 to the nearest hundredth, observe that the hundredths digit is 4 and the thousandths digit is 2. Since 2 is less than 5, the hundredths digit remains unchanged. Therefore, the rounded value is

$$7.14.$$

4. **Solution:**
Begin by expressing each decimal with an equal number of digits after the decimal point. Rewrite 2.5 as 2.500, 2.45 as 2.450, and leave 2.405 as is. Now compare them digit by digit. The whole number digits (2) and the tenths digits are the same for all three. Looking at the hundredths digits, 2.405 has 0, 2.450 has 4, and 2.500 has 5. Since $0 < 4 < 5$, the order from least to greatest is:

$$2.405, \quad 2.450, \quad 2.500.$$

5. **Solution:**
The decimals
$$3.4, \quad 3.40, \quad 3.400$$
are equivalent because appending zeros after the last nonzero digit does not change the value of a number. When normalized, all three represent the same number. Hence,

$$3.4 = 3.40 = 3.400.$$

6. **Solution:**
To round 5.6789 to the nearest hundredth, first identify the hundredths digit, which is 7, and then examine the thousandths digit, which is 8. Since 8 is greater than or equal to

5, increase the hundredths digit by one. Therefore, 5.6789 rounded to the nearest hundredth is

$$5.68.$$

In summary, the rounding process involves identifying the target digit (the hundredths place), checking the next digit (the thousandths place) to decide whether to round up, and adjusting the target digit accordingly.

Chapter 15

Operations with Decimals: Addition and Subtraction

Decimal Place Value and Alignment in Operations

In decimal arithmetic, each digit has a specific place value determined by its position relative to the decimal point. The digit immediately to the left of the decimal represents whole units, while the digits to the right represent fractional values such as tenths, hundredths, thousandths, and so forth. A systematic procedure involves rewriting decimals so that they display an equal number of digits after the decimal point. This normalization, achieved by appending zeros as necessary, guarantees that each corresponding column accurately reflects the same place value. The decimal points must be precisely aligned when the numbers are written one above the other. This arrangement ensures that the operations performed—whether addition or subtraction—proceed with each column representing consistent units of value and prevents incorrect computations that could arise from misaligned digits.

Addition of Decimals

The technique for adding decimals mirrors the common algorithm used for whole numbers, with additional attention given to the decimal point. Once the numbers have been rewritten with the same number of digits after the decimal and lined up vertically by their decimal points, the process commences from the rightmost column and proceeds leftward. Each column is added independently, and if the sum of a column exceeds nine, the excess is carried over to the next column on the left. The decimal point in the result is placed directly below the aligned decimal points of the numbers being added.

1 Step-by-Step Process for Adding Decimals

First, the decimals are rewritten so that each has an equal number of digits after the decimal. For example, consider the numbers 3.45 and 2.78. The number 3.45 may be expressed as 3.450, ensuring that both numbers have three digits after the decimal point. Next, the numbers are arranged in vertical alignment:

$$\begin{array}{r} 3.450 \\ + \ 2.780 \\ \hline \end{array}$$

The addition is performed column by column from right to left. In the hundredths column, the digits are added, and any sum that exceeds nine results in a carried value that must be added to the next higher place column. The decimal point in the answer is placed in the same vertical position as in the numbers being added.

2 Worked Example: Addition Operation

Consider the following configuration:

$$\begin{array}{r} 4.325 \\ + \ 3.678 \\ \hline 8.003 \end{array}$$

In this example, the digits in the thousandths, hundredths, and tenths columns are added sequentially. The procedure includes carrying over when the sum in a given column is greater than nine. The decimal point in the answer is carefully placed to maintain the corresponding place value positions.

Subtraction of Decimals

Subtraction of decimals is executed using a method analogous to the addition process, with the same initial attention given to decimal alignment. The numbers involved are first normalized by ensuring an equal number of digits follow the decimal. Once aligned, the subtraction proceeds from right to left, with each pair of digits being subtracted from one another. In cases where the digit in the minuend is smaller than the corresponding digit in the subtrahend, a regrouping (or borrowing) is performed from the next highest place value. This borrowing adjusts the minuend digit by adding an amount equivalent to the base unit (ten in the context of a single decimal digit), allowing for the subtraction to be carried out correctly in that column.

1 Step-by-Step Process for Subtracting Decimals

The process begins with rewriting both numbers so that they have the same number of digits following the decimal point. For instance, when subtracting 5.60 and 3.48, it may be beneficial to express 5.60 as 5.600 and 3.48 as 3.480. The numbers are then arranged vertically, with decimal points aligned:

$$\begin{array}{r} 5.600 \\ -\ 3.480 \\ \hline \end{array}$$

Subtraction is then carried out column by column. When the digit in a particular column of the top number (minuend) is smaller than the corresponding digit in the bottom number (subtrahend), regrouping is performed by borrowing one unit from the column to the left. This adjustment ensures that the subtraction in that column results in a non-negative number. The decimal point in the difference is placed directly below the decimal points of the original numbers.

2 Worked Example: Subtraction Operation

An illustrative example of subtraction is shown below:

$$\begin{array}{r} 8.650 \\ -\ 3.480 \\ \hline 5.170 \end{array}$$

In this operation, careful attention is given to each column starting with the smallest place value. If the subtraction in any column requires borrowing, the process is executed by reducing the digit in the next higher place value accordingly. Such systematic processing preserves the integrity of each digit's place value and ensures that the computed difference is accurate.

Ensuring Accuracy in Decimal Operations

Accuracy in both addition and subtraction of decimals is deeply rooted in the recognition and management of place values. The systematic normalization of decimals, through the appending of zeros to create uniformity in the number of digits after the decimal, is critical. This standardization permits a precise digit-by-digit operation that adheres to the fundamental rules of arithmetic. Whether carrying over in addition or regrouping in subtraction, every adjustment is made in accordance with the base-ten system, providing clarity and precision in every calculation. Each operation reinforces the understanding that even seemingly small differences in alignment and place value can have substantial effects on the overall computation.

The detailed procedures described illustrate a comprehensive method for performing addition and subtraction operations with decimals. The careful attention to alignment, the diligent application of carrying and borrowing techniques, and the maintenance of consistent place value throughout each step reinforce the critical principles that support accurate and reliable decimal arithmetic.

Multiple Choice Questions

1. When performing operations with decimals, why is it important to align the decimal points?

    (a) It makes the numbers easier to read.
    (b) It ensures that digits in the same place value (units, tenths, hundredths, etc.) are correctly lined up.
    (c) It allows you to add zeros only at the end of the number.
    (d) It changes the value of the numbers.

2. Before adding or subtracting decimals, why do we rewrite the numbers so they have the same number of digits after the decimal point?

   (a) To increase the value of the numbers.

   (b) To simplify the visual alignment, ensuring that every column represents the same place value.

   (c) To convert them into fractions.

   (d) To reduce the number of digits in the calculation.

3. In the addition of decimals, if the sum of a column is 15, what is the proper procedure?

   (a) Write down 15 in the answer's column.

   (b) Carry over 1 to the next column on the left and write 5 in the current column.

   (c) Borrow 1 from the previous column and write 5.

   (d) Multiply 15 by 10 before writing the result.

4. What does "carrying over" mean when adding decimals?

   (a) Transferring an extra value from one column to the next higher place value.

   (b) Writing the entire sum of the column in the answer.

   (c) Ignoring the extra value when the column sum exceeds 9.

   (d) Adjusting the decimal point to the left.

5. During subtraction of decimals, if a digit in the minuend is smaller than the corresponding digit in the subtrahend, what should you do?

   (a) Subtract normally.

   (b) Place a zero in the answer's column.

   (c) Regroup (borrow) from the next column on the left.

   (d) Rearrange the numbers so the larger number is always on top.

6. Which of the following outlines the correct step-by-step process for subtracting decimals?

   (a) Write the numbers in any order and subtract.

(b) Multiply both numbers by 10, then subtract.

(c) Align the decimal points, rewrite the numbers with the same number of digits after the decimal, then subtract column by column from right to left.

(d) Add extra zeros at the beginning of the numbers and then subtract.

7. Why is maintaining accurate place value crucial when performing decimal operations?

(a) It makes the numbers look uniform.

(b) It ensures that each digit contributes the correct value, leading to an accurate overall result.

(c) It speeds up the calculation process.

(d) It eliminates the need for borrowing or carrying.

Answers:

1. **B:** Aligning the decimal points ensures that each digit is in the correct column (units, tenths, hundredths, etc.), which is essential for correctly adding or subtracting corresponding place values.

2. **B:** Rewriting decimals so they have the same number of digits after the decimal point (by appending zeros if needed) simplifies the alignment process. This guarantees that each column represents the same place value, thereby reducing errors.

3. **B:** When the sum of a column exceeds 9 (for example, 15), you write down the ones digit (5) in that column and carry over the extra value (1) to the next column on the left, just as in whole number addition.

4. **A:** "Carrying over" means transferring any value above 9 in a given column to the next higher place value column. This step is critical to maintain accuracy in the sum.

5. **C:** In subtraction, if a digit in the minuend is smaller than the corresponding digit in the subtrahend, you regroup (borrow) from the next higher place value to allow for a proper subtraction in that column.

6. **C:** The correct process is to first align the decimal points and rewrite the numbers so that they have equal digits after the decimal point. Then, subtract digit by digit from right to left, borrowing when necessary.

7. **B:** Maintaining accurate place value is crucial because it ensures that each digit is given its proper weight (units, tenths, hundredths, etc.). This careful attention prevents mistakes and leads to a correct final result.

Practice Problems

1. Add the following decimals by first rewriting them to have the same number of digits after the decimal point and then aligning the decimal points:

$$6.7 \quad \text{and} \quad 3.45$$

2. Subtract the following decimals by ensuring proper alignment and applying the regrouping method when necessary:

$$12.5 - 4.78$$

3. Add the three decimals below by rewriting each so that they have the same number of digits after the decimal point:

$$8.9 + 7.45 + 3.2$$

4. Subtract the following decimals by aligning them and using borrowing where needed:

$$9.065 - 2.38$$

5. A board is 15.3 meters long. If a piece measuring 7.85 meters is cut from the board, what is the length of the remaining piece? Solve by performing the subtraction:

$$15.3 - 7.85$$

6. A student has 20.00 dollars. They spend 7.95 dollars on lunch and 4.25 dollars on a snack. What amount of money do they have left? Solve using decimal operations:

$$20.00 - 7.95 - 4.25$$

Answers

1. **Solution:** To add 6.7 and 3.45, first rewrite 6.7 as 6.70 so that both numbers have two digits after the decimal:

$$\begin{array}{r} 6.70 \\ +\ 3.45 \\ \hline 10.15 \end{array}$$

   Explanation: Aligning the decimal points by rewriting 6.7 as 6.70 ensures that the tenths and hundredths are in the correct columns. Adding column by column gives 0+5 in the hundredths, 7+4 in the tenths (with any carry if necessary), and then the whole numbers.

2. **Solution:** Rewrite 12.5 as 12.50 in order to match the number of decimal places with 4.78:

$$\begin{array}{r} 12.50 \\ -\ 4.78 \\ \hline 7.72 \end{array}$$

   Explanation: Starting from the hundredths place, 0 minus 8 is not possible without borrowing. Borrow 1 from the tenths place, then perform the subtraction digit by digit. This careful regrouping yields the correct result of 7.72.

3. **Solution:** First, rewrite the decimals so that they each show two digits after the decimal:

$$8.9 \to 8.90, \quad 7.45 \text{ remains the same}, \quad 3.2 \to 3.20.$$

Now, add them:
$$\begin{array}{r} 8.90 \\ +\,7.45 \\ \hline 16.35 \\ +\,3.20 \\ \hline 19.55 \end{array}$$

Explanation: Rewriting each number to have the same number of digits after the decimal simplifies the addition. First, add 8.90 and 7.45 to get 16.35, then add 3.20 to arrive at a final answer of 19.55.

4. **Solution:** Write 2.38 as 2.380 so that it has the same number of digits as 9.065:
$$\begin{array}{r} 9.065 \\ -\,2.380 \\ \hline 6.685 \end{array}$$

Explanation: After aligning the decimals, subtract from right to left. In the hundredths column, if a digit in the minuend is smaller than the corresponding digit in the subtrahend, borrow from the higher place value. Following this procedure carefully gives a result of 6.685.

5. **Solution:** Express 15.3 as 15.30:
$$\begin{array}{r} 15.30 \\ -\,7.85 \\ \hline 7.45 \end{array}$$

Explanation: Aligning the decimal points by writing 15.3 as 15.30, subtract starting with the hundredths column. If the digit in a place is too small, borrow from the next column. This process results in a remaining length of 7.45 meters.

6. **Solution:** First, add the amounts spent:

$$7.95 + 4.25 = 12.20$$

Then, subtract this total from 20.00:

$$\begin{array}{r} 20.00 \\ -\ 12.20 \\ \hline 7.80 \end{array}$$

Explanation: Begin by finding the total expenditure by adding 7.95 and 4.25, which gives 12.20. Then subtract 12.20 from 20.00, ensuring proper alignment of the decimal points, to determine that the student has 7.80 dollars remaining.

Chapter 16

Operations with Decimals: Multiplication and Division

Multiplication of Decimals

1 Procedure for Multiplying Decimals

When multiplying decimal numbers, the multiplication proceeds by first ignoring the decimal points and treating the numbers as whole numbers. The multiplication is carried out using the standard algorithm employed for whole numbers. The key aspect in this procedure is to record the number of digits that appear to the right of the decimal in each factor. The total number of decimal places in the product is the sum of these individual counts. This process ensures that the value represented by the decimal point in the final answer reflects the correct order of magnitude consistent with the base-ten system. In calculations, the numbers are multiplied as if they were integers, and once the product is obtained, the decimal point is inserted appropriately by counting from the right end of the number.

2 Worked Example: Multiplying Decimals

Consider the multiplication of two numbers, 4.23 and 1.5. In the first step, the decimal points are removed to treat the numbers as 423 and 15. The product of these whole numbers is computed as follows:
$$423 \times 15 = 6345.$$

The next step requires determining the total number of decimal places present in the factors. The number 4.23 has two digits to the right of the decimal point, and 1.5 has one digit. Together, there are $2 + 1 = 3$ decimal places. The decimal point in the product is then placed so that the resulting number has three digits following it. This adjustment yields the final result:

$$6.345.$$

This method emphasizes the importance of tracking place values and carefully reinstating the decimal point after the multiplication of the whole-number equivalents.

Division of Decimals

1 Procedure for Dividing Decimals

Dividing decimals involves a technique that begins with adjusting the dividend and the divisor so that the divisor becomes a whole number. This is accomplished by multiplying both the dividend and the divisor by the same power of ten. The selection of the appropriate power of ten is determined by the number of digits to the right of the decimal in the divisor. Once the divisor has been converted into a whole number, the division can be performed using the standard algorithm for whole numbers. After the division is complete, the quotient that results from this adjusted operation is equivalent in value to the quotient of the original decimals. This process of eliminating the decimal from the divisor facilitates a straightforward application of the division algorithm while preserving the fundamental ratio between the two numbers.

2 Worked Example: Dividing Decimals

For an illustrative example, consider the division of 7.2 by 0.8. The divisor, 0.8, contains one digit after the decimal point. To convert the divisor into a whole number, both 7.2 (the dividend) and 0.8 (the divisor) are multiplied by 10:

$$7.2 \times 10 = 72 \quad \text{and} \quad 0.8 \times 10 = 8.$$

The division then takes the form:

$$72 \div 8 = 9.$$

In another example with more digits, consider dividing 5.76 by 0.48. The divisor 0.48 has two digits after the decimal point. Multiplying both the dividend and the divisor by 100 yields:

$$5.76 \times 100 = 576 \quad \text{and} \quad 0.48 \times 100 = 48.$$

Carrying out the division:

$$576 \div 48 = 12.$$

This procedure highlights that scaling both numbers with the same factor does not alter the quotient, while it converts the operation into one that involves division of whole numbers. Such adjustment of factors is critical for maintaining the integrity of the calculation and for ensuring that the positional values of the digits are properly accounted for throughout the division process.

Multiple Choice Questions

1. What is the first step when multiplying decimal numbers?

    (a) Multiply the numbers directly while keeping the decimals in place.

    (b) Remove the decimal points and treat the numbers as whole numbers.

    (c) Round the decimals to the nearest whole number before multiplying.

    (d) Multiply only the digits to the left of the decimal points.

2. When multiplying 4.23 by 1.5, how many total decimal places should appear in the final product?

    (a) 2

    (b) 3

(c) 4

(d) 1

3. After multiplying the whole-number equivalents of decimals, how do you correctly place the decimal point in the final answer?

    (a) Place the decimal point in the same position as in the larger factor.

    (b) Place the decimal point exactly in the middle of the product.

    (c) Count the total number of digits to the right of the decimals in the factors and place the decimal point that many digits from the right in the product.

    (d) Place the decimal point at the beginning of the product.

4. In a division problem where the divisor is a decimal, what should you do to simplify the operation?

    (a) Multiply only the dividend by a suitable power of ten.

    (b) Multiply both the dividend and the divisor by the same power of ten to make the divisor a whole number.

    (c) Ignore the decimal portion of the divisor and divide directly.

    (d) Subtract the decimal from both the dividend and the divisor.

5. In the division example 7.2 ÷ 0.8, which process yields the correct whole-number division?

    (a) Multiply both 7.2 and 0.8 by 10 to obtain 72 and 8, then divide.

    (b) Multiply only 7.2 by 10 while keeping 0.8 unchanged.

    (c) Divide 7.2 directly by 0.8 without any modification.

    (d) Multiply both numbers by 100.

6. Why is it important to multiply both the dividend and divisor by the same power of ten when dividing decimals?

    (a) It changes the quotient to a simpler fraction.

    (b) It preserves the original ratio between the numbers and simplifies the divisor to a whole number.

    (c) It only simplifies the dividend.

    (d) It makes the process of division more challenging.

7. When dividing 5.76 by 0.48, why do you multiply both numbers by 100?

    (a) To simplify the dividend only.

    (b) Because 0.48 has two digits after the decimal, multiplying by 100 converts it into a whole number.

    (c) To use a standard multiplication factor, regardless of the numbers.

    (d) To eliminate decimals from the dividend entirely.

Answers:

1. **B:** When multiplying decimals, the first step is to ignore the decimal points and treat the numbers as whole numbers. This allows you to use standard multiplication techniques before reintroducing the decimals into the final product.

2. **B:** The number 4.23 has 2 decimal places and 1.5 has 1 decimal place. The total number of decimal places in the product is $2 + 1 = 3$.

3. **C:** After multiplying the whole-number forms of the decimals, you count the total number of decimal digits in the factors and then position the decimal point in the product that many digits from the right.

4. **B:** Multiplying both the dividend and the divisor by the same power of ten converts the divisor into a whole number, making the division process simpler without changing the quotient.

5. **A:** Multiplying both 7.2 and 0.8 by 10 changes them to 72 and 8 respectively, thus converting the division into a simpler whole-number division ($72 \div 8$).

6. **B:** Using the same factor for both numbers preserves the original ratio between the dividend and divisor. This technique transforms the divisor into a whole number, easing the division process while keeping the quotient unchanged.

7. **B:** Since 0.48 has two digits after the decimal point, multiplying both numbers by 100 converts 0.48 into 48, a whole number. This makes the division straightforward while ensuring accurate placement of the decimal in the final quotient.

Practice Problems

1. Multiply the decimals:

$$3.4 \times 2.6$$

2. Multiply the decimals:

$$0.56 \times 3.2$$

3. Divide the decimals (use the scaling method):

$$7.2 \div 0.4$$

4. Divide the decimals:
$$15.6 \div 1.2$$

5. A pencil costs
$$1.25$$
dollars. If you buy 4 pencils, use decimal multiplication to find the total cost.

6. A container holds
$$0.48$$
liters of water per serving. If you pour
$$5.76$$
liters evenly into servings, how many servings do you get?

Answers

1. For
$$3.4 \times 2.6,$$
we begin by ignoring the decimal points and treat 3.4 as 34 and 2.6 as 26. Multiplying these whole numbers:
$$34 \times 26 = 884.$$
Next, we count the decimal places in the original factors. The number 3.4 has one decimal place, and 2.6 has one decimal place, which means there are a total of $1 + 1 = 2$ decimal places. Inserting the decimal point two digits from the right in 884, we obtain:
$$8.84.$$
Therefore,
$$3.4 \times 2.6 = 8.84.$$

2. For
$$0.56 \times 3.2,$$
we first remove the decimals to consider 0.56 as 56 and 3.2 as 32. Multiplying these whole numbers yields:
$$56 \times 32 = 1792.$$
Now, count the total number of digits to the right of the decimal in the original numbers. Since 0.56 has two decimal places and 3.2 has one, there are $2 + 1 = 3$ decimal places in total. Placing the decimal point in 1792 so that the product has three digits to the right gives:
$$1.792.$$
Thus,
$$0.56 \times 3.2 = 1.792.$$

3. For the division
$$7.2 \div 0.4,$$
we use the method of scaling to convert the divisor into a whole number. Since 0.4 has one digit after the decimal, multiply both the dividend and the divisor by 10:
$$7.2 \times 10 = 72 \quad \text{and} \quad 0.4 \times 10 = 4.$$

Now, divide:
$$72 \div 4 = 18.$$
Therefore,
$$7.2 \div 0.4 = 18.$$

4. For the division
$$15.6 \div 1.2,$$
we again use the scaling method. Because 1.2 has one digit after the decimal, multiply both the dividend and the divisor by 10:
$$15.6 \times 10 = 156 \quad \text{and} \quad 1.2 \times 10 = 12.$$
Now perform the division:
$$156 \div 12 = 13.$$
Hence,
$$15.6 \div 1.2 = 13.$$

5. To find the total cost of 4 pencils when each pencil costs
$$1.25 \text{ dollars},$$
multiply:
$$1.25 \times 4.$$
Think of 1.25 as 125 (ignoring the decimal) and multiplying by 4 gives 500. Since 1.25 has two digits to the right of the decimal, we place the decimal in 500 to yield:
$$5.00 \text{ dollars}.$$
Therefore, the total cost is
$$5.00 \text{ dollars}.$$

6. To determine how many servings you get from pouring
$$5.76 \text{ liters}$$
of water into servings of
$$0.48 \text{ liters per serving},$$

divide the total liters by the amount per serving:
$$5.76 \div 0.48.$$

Multiply both numbers by 100 (since 0.48 has two decimal places) to eliminate the decimals:
$$5.76 \times 100 = 576 \quad \text{and} \quad 0.48 \times 100 = 48.$$

Then, divide:
$$576 \div 48 = 12.$$

Thus, you obtain 12 servings.

Chapter 17

Understanding Percents

Definition of Percents

Percents serve as a means to represent a part of a whole in a standardized form, where the whole is divided into 100 equal segments. In this system, any quantity expressed as a percent indicates how many parts out of 100 are being considered. The notation is derived from the Latin phrase "per centum," meaning "by the hundred," and it is universally denoted by the symbol %. This representation allows for uniform comparisons between different quantities, as every value is scaled relative to one hundred.

Relationship of Percents to Fractions

Every percent can be expressed as a fraction with a denominator of 100. For instance, a percent such as 37% is equivalent to the fraction
$$\frac{37}{100}.$$
In many cases, this fraction can be simplified if the numerator and the denominator have common factors. In the example given, although 37 and 100 do not share any common factors besides 1, other percentages such as 50% yield
$$\frac{50}{100} = \frac{1}{2},$$

after simplification. This equivalence encourages the understanding of percents as a specific type of fraction, providing a bridge between the concepts of proportional parts and classical fraction arithmetic.

Relationship of Percents to Decimals

A percent can also be converted to a decimal by dividing the percent value by 100. This process relocates the decimal point two places to the left. For example, converting 25% to its decimal form involves the operation

$$25 \div 100 = 0.25.$$

Similarly, a percent like 12.5% is expressed in decimal notation as

$$12.5 \div 100 = 0.125.$$

Such transformations demonstrate a direct numerical relationship between percents and decimals, wherein both forms represent the same quantity but are formatted differently for various computational and comparative purposes.

Conversion Techniques

The ability to switch between percents, fractions, and decimals is a valuable skill in many mathematical contexts. Converting from a percent to a fraction begins with the understanding that the percent value is the numerator and 100 is the denominator. This fraction may be simplified by dividing both the numerator and denominator by their greatest common divisor. For example, converting 80% yields

$$\frac{80}{100} = \frac{4}{5},$$

after dividing numerator and denominator by 20.

Similarly, converting a percent to a decimal requires dividing the percent by 100. In doing so, the decimal point in the original number is shifted two positions to the left. For instance, converting 75% results in

$$75 \div 100 = 0.75.$$

When converting in the reverse direction, a fraction or decimal is transformed into a percent by means of multiplication by 100. If a decimal such as 0.6 is presented, multiplying by 100 results in

$$0.6 \times 100 = 60\%,$$

while a fraction such as

$$\frac{3}{4}$$

can be converted by first expressing it in decimal form (0.75) and then multiplying by 100 to obtain 75%.

Examples of Conversions

Consider the percent 60%. Expressing it as a fraction involves writing

$$\frac{60}{100},$$

which can be simplified by dividing both numbers by 20 to yield

$$\frac{3}{5}.$$

The decimal form is achieved by the division

$$60 \div 100 = 0.60.$$

Another instance involves the percent 12.5%. As a fraction, this value is initially written as

$$\frac{12.5}{100}.$$

Multiplying the numerator and the denominator by 10 to eliminate the decimal produces

$$\frac{125}{1000},$$

which simplifies to

$$\frac{1}{8},$$

upon dividing both terms by 125. Converting 12.5% to decimal form is direct, yielding

$$12.5 \div 100 = 0.125.$$

These examples illustrate the procedures for transitioning seamlessly between the different representations of a number. The consistent use of a common base, in this case 100, ensures that relationships among the forms remain clear and logical, reinforcing the fundamental concept that percents articulate parts of a hundred.

Multiple Choice Questions

1. Which of the following best defines a percent?

    (a) A fraction with a denominator of 10.

    (b) A part of a whole expressed out of 100.

    (c) A decimal number rounded to two places.

    (d) A ratio comparing one quantity to another with no fixed total.

2. What is the equivalent simplified fraction for 40%?

    (a) $\dfrac{2}{5}$

    (b) $\dfrac{4}{10}$

    (c) $\dfrac{1}{4}$

    (d) $\dfrac{40}{1}$

3. How can 75% be expressed as a decimal?

    (a) 0.075

    (b) 0.75

    (c) 7.5

    (d) 75.0

4. What is the correct process to convert a percent to a fraction in simplest form?

    (a) Write the percent value as the numerator over 100 and then simplify.

    (b) Multiply the percent value by 100 and simplify.

    (c) Divide the percent value by 100 and then add one.

(d) Express the percent as a decimal and then convert it to a fraction.

5. 12.5% is equivalent to which of the following fractions in simplest form?

   (a) $\dfrac{1}{8}$

   (b) $\dfrac{5}{8}$

   (c) $\dfrac{12.5}{100}$

   (d) $\dfrac{1}{12.5}$

6. To convert a fraction to a percent, what is the correct procedure using the example $\dfrac{3}{4}$?

   (a) Multiply 3 by 4 and then multiply by 100.

   (b) Divide 3 by 4 to obtain a decimal, then multiply the result by 100.

   (c) Divide 4 by 3 to obtain a decimal, then multiply the result by 100.

   (d) Multiply 3 by 100 and then divide by 4.

7. Which of the following conversions for 80% is correct?

   (a) $\dfrac{80}{100} = \dfrac{4}{5}$ and $80\% = 0.8$

   (b) $\dfrac{80}{100} = \dfrac{8}{10}$ and $80\% = 0.08$

   (c) $\dfrac{80}{100} = \dfrac{4}{5}$ and $80\% = 8$

   (d) $\dfrac{80}{100} = \dfrac{1}{2}$ and $80\% = 0.8$

Answers:

1. **B: A part of a whole expressed out of 100**
   Explanation: The term "percent" comes from the Latin "per centum," meaning "by the hundred." Thus, a percent represents a ratio or fraction with a denominator of 100.

2. **A:** $\frac{2}{5}$

    Explanation: Expressing 40% as a fraction gives $\frac{40}{100}$. When you simplify $\frac{40}{100}$ by dividing both the numerator and denominator by 20, you obtain $\frac{2}{5}$.

3. **B: 0.75**

    Explanation: To convert 75% to a decimal, divide 75 by 100, which results in 0.75. This shifts the decimal point two places to the left.

4. **A: Write the percent value as the numerator over 100 and then simplify**

    Explanation: Converting a percent to a fraction involves writing it as $\frac{\text{percent value}}{100}$ and then reducing the fraction to its simplest form if possible.

5. **A:** $\frac{1}{8}$

    Explanation: Starting with 12.5% written as $\frac{12.5}{100}$, multiply numerator and denominator by 10 to eliminate the decimal to get $\frac{125}{1000}$; then simplifying by dividing numerator and denominator by 125 yields $\frac{1}{8}$.

6. **B: Divide 3 by 4 to obtain a decimal, then multiply the result by 100**

    Explanation: To convert a fraction like $\frac{3}{4}$ to a percent, first divide 3 by 4 to get 0.75, and then multiply 0.75 by 100 to arrive at 75%.

7. **A:** $\frac{80}{100} = \frac{4}{5}$ **and** $80\% = 0.8$

    Explanation: 80% expressed as a fraction is $\frac{80}{100}$ which simplifies to $\frac{4}{5}$. Additionally, converting 80% to a decimal involves dividing 80 by 100, which gives 0.8.

Practice Problems

1. Convert 50% into a fraction in its simplest form and then express it as a decimal.

2. Express 12.5% as a fraction in simplest form and as a decimal.

3. Convert the fraction
$$\frac{3}{5}$$
into a percent.

4. If 75% of a number is 45, find the original number.

5. Convert 80% into a fraction (simplified) and a decimal.

6. A store offers a 25% discount on a jacket originally priced at 80. Determine the discount amount, the sale price, and express the discount as a fraction and as a decimal.

Answers

1. **Solution:** By definition, 50% means 50 out of 100. Thus, we write
$$50\% = \frac{50}{100}.$$
To simplify, divide the numerator and denominator by 50:
$$\frac{50 \div 50}{100 \div 50} = \frac{1}{2}.$$
Next, to express $\frac{1}{2}$ as a decimal, divide 1 by 2:
$$\frac{1}{2} = 0.5.$$
Therefore, 50% is equivalent to $\frac{1}{2}$ and 0.5.

2. **Solution:** To express 12.5% as a fraction, write
$$12.5\% = \frac{12.5}{100}.$$
To eliminate the decimal in the numerator, multiply both numerator and denominator by 10:
$$\frac{12.5 \times 10}{100 \times 10} = \frac{125}{1000}.$$
Now simplify by dividing both the numerator and denominator by 125:
$$\frac{125 \div 125}{1000 \div 125} = \frac{1}{8}.$$
For the decimal form, simply divide 12.5 by 100:
$$12.5 \div 100 = 0.125.$$
Hence, 12.5% is equivalent to $\frac{1}{8}$ and 0.125.

3. **Solution:** To convert the fraction
$$\frac{3}{5}$$
into a percent, first express it as a decimal by dividing 3 by 5:
$$\frac{3}{5} = 0.6.$$
Then, multiply by 100 to convert the decimal into a percent:
$$0.6 \times 100 = 60\%.$$
Therefore, $\frac{3}{5}$ is equivalent to 60%.

4. **Solution:** Let the unknown number be N. We are told that 75% of N is 45. Write this as
$$0.75 \times N = 45.$$
To solve for N, divide both sides by 0.75:
$$N = \frac{45}{0.75}.$$
Performing the division:
$$\frac{45}{0.75} = 60.$$
Hence, the original number is 60.

5. **Solution:** Start by converting 80% to a fraction:
$$80\% = \frac{80}{100}.$$
Simplify by dividing the numerator and denominator by 20:
$$\frac{80 \div 20}{100 \div 20} = \frac{4}{5}.$$
Next, convert 80% to a decimal by dividing 80 by 100:
$$80 \div 100 = 0.8.$$
Thus, 80% is equivalent to $\frac{4}{5}$ as a fraction and 0.8 as a decimal.

6. **Solution:** The original price of the jacket is 80. A 25% discount means that the discount amount is 25% of 80. First, express 25% as a decimal:
$$25\% = 0.25.$$
Now, calculate the discount amount:
$$0.25 \times 80 = 20.$$
The sale price is determined by subtracting the discount from the original price:
$$80 - 20 = 60.$$
Next, express 25% as a fraction:
$$25\% = \frac{25}{100}.$$
Simplify by dividing the numerator and denominator by 25:
$$\frac{25 \div 25}{100 \div 25} = \frac{1}{4}.$$
The decimal form of the discount percentage is already known as 0.25.

In summary, the discount amount is 20, the sale price is 60, and the discount can be expressed as $\frac{1}{4}$ or 0.25.

Chapter 18

Percent Problems and Conversions

Converting Percents to Decimals and Fractions

Percent values represent parts per one hundred and can be transformed into decimals or fractions through systematic procedures. A percent is always expressed out of one hundred. Converting a percent to a fraction involves writing the numerical value of the percent as the numerator and 100 as the denominator. For instance, a percent such as 37% becomes

$$\frac{37}{100}.$$

In cases where the numerator and denominator share common factors, the fraction can be simplified; for example, 50% is represented by

$$\frac{50}{100},$$

which simplifies to

$$\frac{1}{2}.$$

The process of converting a percent to a decimal is accomplished by dividing the percent value by 100, effectively shifting the decimal point two places to the left. Thus, converting 37% yields

$$37 \div 100 = 0.37,$$

and converting 50% gives
$$50 \div 100 = 0.50.$$

Converting Decimals and Fractions to Percents

The reverse conversion requires multiplying a decimal by 100 to obtain the percent form. A decimal such as 0.75, when multiplied by 100, results in
$$0.75 \times 100 = 75\%.$$
For fractions, the conversion is performed indirectly by first converting the fraction to its decimal representation through division of the numerator by the denominator. As an example, the fraction
$$\frac{3}{4}$$
divides to yield 0.75. Multiplying this decimal value by 100 transforms it into a percent:
$$0.75 \times 100 = 75\%.$$
Such conversion techniques provide a clear framework to interpret and compare values, whether they are presented as fractions, decimals, or percents.

Solving Percent Problems in Applied Situations

Problems involving percents often require the determination of a part, a whole, or the percent value itself. When the part and the percent are known, the original whole can be determined by setting up an equation. For example, if 75% of an unknown quantity is 45, the equation
$$0.75 \times N = 45$$
can be solved by isolating N as follows:
$$N = \frac{45}{0.75},$$

which evaluates to 60. This method demonstrates the use of percent-to-decimal conversion in an algebraic context.

Another application involves computing discounts or percentage decreases. Consider an item with an original price of 80 that is reduced by 25%. Converting 25% to a decimal gives

$$25 \div 100 = 0.25.$$

The discount amount is computed by multiplying the original price by this decimal:
$$0.25 \times 80 = 20.$$

Subtracting the discount from the original price yields the final sale price:
$$80 - 20 = 60.$$

These calculations illustrate how percent conversions are instrumental in solving practical problems and making quantitative comparisons.

Worked Examples and Practice Problems

To illustrate the concepts further, consider the conversion of 80% into both fraction and decimal forms. Start by expressing 80% as a fraction:
$$\frac{80}{100}.$$

Dividing both numerator and denominator by 20 simplifies the fraction to:
$$\frac{4}{5}.$$

Next, converting 80% to a decimal involves dividing 80 by 100, resulting in:
$$80 \div 100 = 0.8.$$

Another detailed example involves reversing the process. Suppose that 30% of an unknown quantity equals 24. By expressing 30% as the decimal 0.30, the corresponding equation is:
$$0.30 \times N = 24.$$

Dividing both sides of the equation by 0.30 results in:
$$N = \frac{24}{0.30} = 80.$$

This approach reinforces the systematic use of decimal manipulation in determining an unknown total.

An additional example concerns the conversion of a fraction into a percent. For the fraction
$$\frac{1}{5},$$
the division 1 ÷ 5 yields 0.2. Multiplying this decimal by 100 produces:
$$0.2 \times 100 = 20\%.$$
This conversion emphasizes the proportionality inherent in the base of 100, allowing for straightforward comparisons among different numeric forms.

The techniques outlined above serve to reinforce the foundational concepts of converting between percents, decimals, and fractions as well as applying these conversions to solve diverse percent-related problems.

Multiple Choice Questions

1. Which of the following is the correct decimal equivalent of 25%?

    (a) 2.5

    (b) 0.25

    (c) 0.025

    (d) 25

2. Which fraction is equivalent to 75% in simplest form?

    (a) $\frac{1}{3}$

    (b) $\frac{3}{4}$

    (c) $\frac{2}{3}$

    (d) $\frac{3}{5}$

3. What is the percent form of the decimal 0.6?

    (a) 6%

    (b) 16%

    (c) 60%

(d) 600%

4. If 30% of an unknown number is 18, what is the number?

    (a) 54
    (b) 60
    (c) 72
    (d) 48

5. How can you simplify the fraction that represents 40%?

    (a) $\frac{40}{100}$ simplifies to $\frac{2}{5}$
    (b) $\frac{40}{100}$ simplifies to $\frac{4}{10}$
    (c) $\frac{40}{100}$ simplifies to $\frac{1}{2}$
    (d) $\frac{40}{100}$ simplifies to $\frac{4}{5}$

6. An item is originally priced at $80 and is discounted by 20%. What is its sale price?

    (a) $64
    (b) $60
    (c) $70
    (d) $75

7. When converting a percent to a decimal, which of the following operations is correct?

    (a) Multiply the percent by 0.1
    (b) Divide the percent value by 100
    (c) Multiply the percent by 100
    (d) Subtract 100 from the percent value

Answers:

1. **B: 0.25**

    To convert 25% to a decimal, divide the number by 100. That is, $25 \div 100 = 0.25$.

2. **B: $\frac{3}{4}$**

    Express 75% as a fraction: $\frac{75}{100}$. Simplify by dividing numerator and denominator by 25, giving $\frac{3}{4}$.

3. **C: 60%**
   To convert the decimal 0.6 to a percent multiply by 100: $0.6 \times 100 = 60\%$.

4. **B: 60**
   The equation $0.30 \times N = 18$ represents 30% of an unknown number N being 18. Solve by dividing: $N = 18 \div 0.30 = 60$.

5. **A: $\frac{40}{100}$ simplifies to $\frac{2}{5}$**
   Write 40% as $\frac{40}{100}$ and simplify the fraction by dividing numerator and denominator by 20, which yields $\frac{2}{5}$.

6. **A: $64**
   A 20% discount on $80 means the discount amount is $0.20 \times 80 = 16$. Subtracting from the original price: $80 - 16 = 64$.

7. **B: Divide the percent value by 100**
   Converting a percent to a decimal requires dividing by 100. For example, converting 25% gives $25 \div 100 = 0.25$.

Practice Problems

1. Convert 25% to a fraction and a decimal.

2. Convert the decimal 0.85 to a percent.

3. If 40% of a number is 32, what is the number?

4. An item originally costs 120. It is discounted by 30%. Find the discount amount and the sale price.

5. Convert the fraction
$$\frac{3}{8}$$
to a percent.

6. If 15% of a number is 24, find the original number.

Answers

1. **Solution:** To convert 25% to a fraction, remember that a percent represents parts out of one hundred. Thus,
$$25\% = \frac{25}{100}.$$
Simplify this fraction by dividing the numerator and denominator by their greatest common divisor, 25:
$$\frac{25 \div 25}{100 \div 25} = \frac{1}{4}.$$
To convert 25% to a decimal, divide 25 by 100:
$$25 \div 100 = 0.25.$$
Therefore, 25% is equivalent to $\frac{1}{4}$ and 0.25.

2. **Solution:** To convert the decimal 0.85 into a percent, multiply by 100 (which shifts the decimal point two places to the right):
$$0.85 \times 100 = 85.$$
Attach the percent symbol:
$$85\%.$$
Hence, 0.85 is equivalent to 85%.

3. **Solution:** Let the unknown number be N. The problem states that 40% of N is 32. This is expressed mathematically as:
$$0.40 \times N = 32.$$
To solve for N, divide both sides by 0.40:
$$N = \frac{32}{0.40} = 80.$$
Therefore, the original number is 80.

4. **Solution:** First, calculate the discount amount by finding 30% of 120:
$$0.30 \times 120 = 36.$$

This means the discount is 36. Next, subtract the discount from the original price to determine the sale price:

$$120 - 36 = 84.$$

Thus, the discount amount is 36 and the sale price is 84.

5. **Solution:** To convert the fraction $\frac{3}{8}$ to a percent, start by converting it to a decimal. Divide 3 by 8:

$$\frac{3}{8} = 0.375.$$

Then multiply the decimal by 100 to change it into a percent:

$$0.375 \times 100 = 37.5.$$

So,
$$\frac{3}{8} = 37.5\%.$$

6. **Solution:** Let the unknown number be N. According to the problem, 15% of N is 24, which can be written as:

$$0.15 \times N = 24.$$

Solve for N by dividing both sides by 0.15:

$$N = \frac{24}{0.15} = 160.$$

Therefore, the original number is 160.

Chapter 19

Ratios and Proportions

Understanding Ratios

A ratio is a mathematical expression that compares two quantities by indicating how many times one quantity is contained in another. Ratios can be represented using a colon, as in 3:4, or as a fraction, such as $\frac{3}{4}$. This representation illustrates the relationship between two measures in a clear format. For example, in a ratio expressed as 5:2, the first quantity is five parts while the second quantity is two parts. The comparison establishes a relative scale that remains independent of any units, provided both quantities are measured on the same scale.

Representing and Simplifying Ratios

Ratios are often simplified to reveal the most basic relation between quantities. The process of simplifying a ratio is analogous to reducing a fraction to its simplest form. When a ratio such as $\frac{8}{12}$ is encountered, both numerator and denominator can be divided by their greatest common divisor, in this case 4, to produce the simplified ratio $\frac{2}{3}$. This simplification not only makes the numerical relationship easier to compare with other ratios but also exposes the underlying proportional relationship between the quantities. Different representations, including colon notation (8:12 simplifying to 2:3) and fractional form, serve the same purpose: to convey the relative sizes of the compared values in the clearest way possible.

Exploring Proportional Relationships

A proportional relationship arises when two ratios are equivalent. In other words, if two ratios express the same comparison between pairs of quantities, they are said to be proportional. This concept is critical for understanding how one quantity changes in direct correspondence with another. When variables vary in such a way that the ratio between them remains constant, the relationship is considered to be directly proportional. The concept can be expressed mathematically in the form $\frac{a}{b} = \frac{c}{d}$, where a, b, c, and d represent quantities that maintain a constant ratio. This constant ratio is often referred to as the unit rate and is a foundational element in the study of proportionality.

Solving Proportions

Solving a proportion involves finding an unknown value that maintains the equivalence between two ratios. The common method for solving such equations is cross multiplication. Given a proportion of the form
$$\frac{a}{b} = \frac{c}{d},$$
cross multiplication leads to the equation
$$a \times d = b \times c.$$
This equation can then be solved for the unknown variable. For instance, if the proportion is expressed as
$$\frac{x}{4} = \frac{6}{8},$$
then cross multiplying yields
$$8x = 4 \times 6.$$
Solving this equation, $8x = 24$ leads to $x = 3$. This procedure demonstrates the systematic approach used to determine the unknown when two ratios are set equal to one another.

Applications of Ratios and Proportions

Ratios and proportions have significant applications in numerous real-life scenarios, ranging from scaling recipes to analyzing maps and models. In visual representations such as maps, a scale ratio might indicate that one unit on the map corresponds to a fixed number of real-world units, ensuring accurate representation of distances. In everyday contexts, ratios may be used to compare ingredients in a recipe, where maintaining the proper proportions is essential to achieving the desired outcome. Additionally, proportional reasoning is a key element in statistical analysis, where quantities such as rates, averages, and percentages are examined to discern patterns and make predictions. The constant relationship between the quantities, as established by proportionality, provides a reliable framework for these diverse applications.

Multiple Choice Questions

1. Which of the following best describes a ratio?

    (a) A ratio is the sum of two quantities.

    (b) A ratio is a comparison of two quantities showing how many times one value contains the other.

    (c) A ratio is the product of two quantities.

    (d) A ratio is the difference between two quantities.

2. When simplifying the ratio 8:12, what is the result?

    (a) 4:6

    (b) 2:3

    (c) 3:2

    (d) 1:2

3. Which equation must hold true for two ratios a:b and c:d to be proportional?

    (a) a + d = b + c

    (b) a - b = c - d

    (c) a × d = b × c

    (d) a/b = c/d only when a = c

4. What is the first step in solving the proportion $\frac{x}{4} = \frac{6}{8}$?

    (a) Adding 4 to both sides.

    (b) Cross multiplying to obtain $8x = 4 \times 6$.

(c) Dividing both sides by 6.

(d) Multiplying both sides by 4.

5. Which strategy is most effective for solving proportional equations?

    (a) Cross multiplication.

    (b) Using the distributive property.

    (c) Adding the numerators and denominators.

    (d) Subtracting the lesser term from the greater.

6. Which of the following scenarios is a real-life application of ratios and proportions?

    (a) Calculating the area of a circle.

    (b) Balancing a chemical equation.

    (c) Scaling a recipe.

    (d) Determining the perimeter of a rectangle.

7. If a map scale indicates that 1 inch represents 5 miles, how many miles does 3 inches represent?

    (a) 3 miles

    (b) 5 miles

    (c) 15 miles

    (d) 20 miles

Answers:

1. **B: A ratio is a comparison of two quantities showing how many times one value contains the other.** Explanation: This definition correctly highlights that a ratio compares the relative amounts of two values, independent of their units.

2. **B: 2:3** Explanation: Both 8 and 12 are divisible by 4, so dividing each term by 4 simplifies the ratio 8:12 to 2:3.

3. **C: a × d = b × c** Explanation: For two ratios to be proportional, their cross products must be equal. This equality is the basis of cross multiplication.

4. **B: Cross multiplying to obtain** $8x = 4 \times 6$. Explanation: Cross multiplying the proportion $\frac{x}{4} = \frac{6}{8}$ gives $8x = 24$, which is the correct first step in solving for x.

5. **A: Cross multiplication.** Explanation: Cross multiplication is the standard and most efficient method for solving equations that set two ratios equal to each other.

6. **C: Scaling a recipe.** Explanation: Using ratios and proportions is essential in real-life tasks like adjusting ingredient amounts in a recipe to maintain correct flavor and consistency.

7. **C: 15 miles** Explanation: With a map scale of 1 inch representing 5 miles, multiplying 3 inches by 5 gives $3 \times 5 = 15$ miles.

Practice Problems

1. Simplify the ratio 8:12. Express your answer in both fractional form and colon notation.

2. Solve for y in the proportion:
$$\frac{8}{y} = \frac{10}{15}$$

3. Determine whether the two ratios 4:5 and 8:10 are proportional. Explain your reasoning.

4. In a recipe, the ratio of butter to sugar is 2:3. If a baker uses 4 cups of butter, how many cups of sugar are needed?

5. A map uses a scale where 1 centimeter represents 20 meters. If the distance between two landmarks on the map is 7 centimeters, what is the actual distance in meters?

6. The student council at a school has a ratio of boys to girls of 5:4. If there are 45 boys, find the number of girls.

Answers

1. **Simplify the Ratio 8:12**
   To simplify the ratio 8:12, first express it as the fraction
   $$\frac{8}{12}.$$
   The greatest common divisor (GCD) of 8 and 12 is 4. Divide both numerator and denominator by 4:
   $$\frac{8 \div 4}{12 \div 4} = \frac{2}{3}.$$
   In colon notation, this simplified ratio is written as 2:3.
   Answer: $\frac{2}{3}$ or 2:3.

2. **Solve for y in the proportion** $\frac{8}{y} = \frac{10}{15}$
   First, simplify the fraction on the right:
   $$\frac{10}{15} = \frac{10 \div 5}{15 \div 5} = \frac{2}{3}.$$
   Now the equation becomes:
   $$\frac{8}{y} = \frac{2}{3}.$$
   Use cross multiplication:
   $$8 \times 3 = 2 \times y.$$
   Simplify:
   $$24 = 2y.$$
   Divide both sides by 2:
   $$y = 12.$$
   Answer: $y = 12$.

3. **Determine if 4:5 and 8:10 are proportional**
   To check if the ratios are proportional, we compare their simplified forms. The ratio 4:5 can also be written as:
   $$\frac{4}{5}.$$

For the second ratio, 8:10, simplify by dividing both terms by 2:
$$\frac{8 \div 2}{10 \div 2} = \frac{4}{5}.$$
Since both ratios simplify to $\frac{4}{5}$, the ratios are proportional.
Answer: Yes; both ratios represent the same relationship.

4. **Butter to Sugar Ratio in a Recipe**
   The ratio of butter to sugar is given as 2:3. This means for every 2 cups of butter, 3 cups of sugar are needed. If 4 cups of butter are used, set up the proportion:
   $$\frac{2}{3} = \frac{4}{s},$$
   where s represents the cups of sugar. Cross multiply:
   $$2 \times s = 3 \times 4.$$

Simplify:
$$2s = 12.$$

Divide both sides by 2:
$$s = 6.$$

Answer: 6 cups of sugar are needed.

5. **Map Scale Problem**
   The map scale indicates 1 centimeter represents 20 meters. If the distance on the map is 7 centimeters, multiply to find the actual distance:
   $$7\,\text{cm} \times 20\,\text{meters/cm} = 140\,\text{meters}.$$

   Answer: 140 meters.

6. **Student Council Ratio Problem**
   The ratio of boys to girls is given as 5:4, and there are 45 boys. Let the number of girls be g. The ratio can be set up as:
   $$\frac{5}{4} = \frac{45}{g}.$$
   Cross multiply:
   $$5g = 4 \times 45.$$

Calculate:
$$5g = 180.$$

Divide both sides by 5:
$$g = 36.$$

Answer: There are 36 girls.

Chapter 20

Solving Proportional Problems

Direct Proportionality and Equivalent Ratios

Direct proportionality describes a relationship between two quantities in which an increase or decrease in one quantity is accompanied by a corresponding increase or decrease in the other. In such relationships the ratio between the quantities remains constant regardless of the scale. This constant ratio is known as the constant of proportionality and is expressed mathematically by an equation of the form

$$A = kB,$$

where A and B are the related quantities and k is the constant of proportionality. In comparing parts of these quantities, equivalent ratios reveal the same relationship. For example, if one ratio is written as

$$\frac{a}{b},$$

any other ratio equivalent to it will simplify to the same fraction, confirming the existence of an unchanging proportional relationship between the quantities involved.

Establishing Proportional Equations

The process of setting up a proportional equation begins with representing the relationship between pairs of quantities as fractions. A typical proportional equation takes the form
$$\frac{a}{b} = \frac{c}{d},$$
where the numerators and denominators correspond to parts of the quantities under comparison. In this context the equality holds true if the product of the means equals the product of the extremes; that is, if
$$a \times d = b \times c.$$
This equality serves as the foundation for verifying that the ratios on both sides represent an equivalent relationship and provides a method to relate an unknown value to known quantities.

Solving Proportions Using Cross Multiplication

The method of cross multiplication transforms a proportional equation into a linear equation that is straightforward to solve. When presented with the proportion
$$\frac{a}{b} = \frac{c}{d},$$
multiplying across the equality produces
$$a \times d = b \times c.$$

This method effectively removes the fractions and isolates the unknown quantity when one of the terms is not provided. The resulting equation can be manipulated using the standard techniques of algebra. Cross multiplication is highly efficient in solving problems that involve direct proportions, as it directly exploits the inherent equality of the two ratios.

Worked Examples

Consider the proportion
$$\frac{x}{6} = \frac{4}{9}.$$

Applying cross multiplication yields
$$9x = 6 \times 4.$$
Simplification leads to
$$9x = 24,$$
and dividing both sides by 9 gives
$$x = \frac{24}{9},$$
which simplifies to
$$x = \frac{8}{3}.$$

In another example, examine the proportion
$$\frac{10}{y} = \frac{15}{20}.$$
The fraction on the right simplifies by dividing the numerator and the denominator by 5, resulting in
$$\frac{15}{20} = \frac{3}{4}.$$
The proportion then becomes
$$\frac{10}{y} = \frac{3}{4}.$$
Cross multiplication yields
$$10 \times 4 = 3 \times y,$$
or
$$40 = 3y.$$
Dividing both sides by 3 produces
$$y = \frac{40}{3}.$$

A practical application can be observed in a map scale scenario. Suppose a scale indicates that 1 centimeter corresponds to 30 miles. With a measured distance of 5 centimeters on the map, a proportion can be set up as
$$\frac{1}{30} = \frac{5}{d},$$

where d represents the actual distance in miles. Cross multiplication results in
$$1 \times d = 30 \times 5,$$
which simplifies to
$$d = 150.$$

Applications to Problem Solving

The techniques of setting up and solving proportional equations have widespread applications in various real-world scenarios. Scaling geometric figures, adjusting quantities in recipes, and interpreting map scales all rely on an understanding of direct proportions. In each instance the constant ratio guarantees that as one quantity changes, the corresponding quantity changes in a consistent manner. The process begins with translating the given situation into an equation where two ratios are equated. Once this proportional equation is established, cross multiplication converts the relationship into a solvable algebraic equation. The method reveals the underlying structure of the problem, demonstrating that the equality of ratios is preserved regardless of the numerical values involved. This approach underscores the elegance and practicality of proportional reasoning in solving everyday problems involving equivalent relationships.

Multiple Choice Questions

1. Which of the following equations best represents a direct proportionality relationship between two quantities A and B?

    (a) A = k + B

    (b) A = kB

    (c) A = k - B

    (d) A = $\dfrac{k}{B}$

2. In a proportional relationship expressed as
$$\frac{a}{b} = \frac{c}{d},$$
which statement is true?

(a) $a + d = b + c$

(b) $a - c = b - d$

(c) $a \times d = b \times c$ $\dfrac{a}{d} = \dfrac{b}{c}$

(3) Solve for x in the proportion
$$\dfrac{3}{x} = \dfrac{6}{8}.$$

(a) $x = 2$

(b) $x = 4$

(c) $x = 6$

(d) $x = 8$

4. What is the first step in solving a proportional problem using cross multiplication?

   (a) Multiply both sides by the unknown variable immediately.

   (b) Set up the relationship as two equal fractions (a proportion).

   (c) Simplify one of the ratios to its lowest terms.

   (d) Substitute the known values after cross multiplying.

5. A map scale indicates that 1 centimeter represents 30 miles. How many miles are represented by 5 centimeters on the map?

   (a) 30 miles

   (b) 150 miles

   (c) 100 miles

   (d) 180 miles

6. Which part of the cross multiplication process eliminates the fractions?

   (a) Multiplying the denominators together.

   (b) Multiplying the numerator of one fraction by the denominator of the other.

   (c) Dividing both sides by the unknown variable.

(d) Subtracting the smaller fraction from the larger.

7. After cross multiplication yields an equation such as
$$9x = 24,$$
what is the proper next step to solve for x?

(a) Add 9 to both sides.

(b) Divide both sides by 9.

(c) Multiply both sides by 9.

(d) Subtract 24 from both sides.

Answers:

1. **B: A = kB**
   In direct proportionality, one quantity is a constant multiple of the other, which is expressed by the equation A = kB where k is the constant of proportionality.

2. **C:** $a \times d = b \times c$

   This equality, known as cross multiplication, confirms that the product of the means equals the product of the extremes in a valid proportion.

3. **B:** $x = 4$

   Cross multiplying gives $3 \times 8 = 6x$, so $24 = 6x$. Dividing both sides by 6 yields $x = 4$.

4. **B: Set up the relationship as two equal fractions (a proportion).**
   The first step is to express the situation as a proportion; once the relationship is written as two equal fractions, cross multiplication can be applied to solve for the unknown.

5. **B: 150 miles**
   Using the scale, multiply the map distance by the conversion factor: 5 cm × 30 miles/cm = 150 miles.

6. **B: Multiplying the numerator of one fraction by the denominator of the other.**
   Cross multiplication involves multiplying across the equal sign (numerator of one fraction by the denominator of the other), which effectively removes the fractional form and yields a linear equation.

7. **B: Divide both sides by 9.**
   Once the equation 9x = 24 is obtained, isolating x requires dividing both sides by 9 to solve for the unknown value.

Practice Problems

1. Given the formula for direct proportionality,
$$A = kB,$$
if
$$A = 12 \quad \text{when} \quad B = 4,$$
find the constant of proportionality k and then calculate the value of A when $B = 10$.

2. Solve the proportion for the unknown x:
$$\frac{x-2}{5} = \frac{3}{10}$$

3. A recipe uses 2 cups of flour to make 24 cookies. How many cups of flour are needed to make 36 cookies? Set up a proportion and solve for the unknown amount of flour.

4. A map has a scale where 1 cm represents 30 miles. If the distance between two cities measures 4.5 cm on the map, what is the actual distance between the cities?

5. Solve for y in the proportion:
$$\frac{7}{y} = \frac{21}{15}$$

6. If 3 notebooks cost 9 dollars, how much will 7 notebooks cost? Use a proportional equation to solve this problem.

Answers

1. **Solution:** We start with the equation
$$A = kB.$$
Given that $A = 12$ when $B = 4$, substitute these values into the equation:
$$12 = k \times 4.$$
Divide both sides by 4 to solve for k:
$$k = \frac{12}{4} = 3.$$
Now, using the constant $k = 3$, find A when $B = 10$:
$$A = 3 \times 10 = 30.$$
Therefore, the constant of proportionality is 3 and $A = 30$ when $B = 10$.

2. **Solution:** The given proportion is
$$\frac{x-2}{5} = \frac{3}{10}.$$
Cross multiply to eliminate the fractions:
$$10(x-2) = 5 \times 3.$$
Simplify:
$$10(x-2) = 15.$$
Distribute on the left side:
$$10x - 20 = 15.$$
Add 20 to both sides:
$$10x = 35.$$
Divide both sides by 10:
$$x = \frac{35}{10} = 3.5.$$
Thus, the solution is $x = 3.5$.

3. **Solution:** Let x be the number of cups of flour needed for 36 cookies. The proportion based on the recipe is:

$$\frac{2 \text{ cups}}{24 \text{ cookies}} = \frac{x \text{ cups}}{36 \text{ cookies}}.$$

Cross multiply:
$$2 \times 36 = 24 \times x.$$

Simplify the multiplication:
$$72 = 24x.$$

Divide both sides by 24:
$$x = \frac{72}{24} = 3.$$

Therefore, 3 cups of flour are needed to make 36 cookies.

4. **Solution:** According to the map scale,
$$1 \text{ cm} = 30 \text{ miles}.$$

Let d be the actual distance corresponding to 4.5 cm on the map. Set up the proportion:

$$\frac{1 \text{ cm}}{30 \text{ miles}} = \frac{4.5 \text{ cm}}{d}.$$

Cross multiply:
$$1 \times d = 30 \times 4.5.$$

Compute the product:
$$d = 135.$$

So the actual distance between the cities is 135 miles.

5. **Solution:** The proportion given is:
$$\frac{7}{y} = \frac{21}{15}.$$

Cross multiply to solve for y:
$$7 \times 15 = 21 \times y.$$

Calculate the multiplication:
$$105 = 21y.$$

Divide both sides by 21:
$$y = \frac{105}{21} = 5.$$

Thus, $y = 5$.

6. **Solution:** Let x represent the cost of 7 notebooks. We are given that 3 notebooks cost 9 dollars. Set up the proportion:
$$\frac{3 \text{ notebooks}}{9 \text{ dollars}} = \frac{7 \text{ notebooks}}{x \text{ dollars}}.$$

Cross multiply:
$$3x = 9 \times 7.$$

Simplify:
$$3x = 63.$$

Divide both sides by 3:
$$x = \frac{63}{3} = 21.$$

Therefore, 7 notebooks cost 21 dollars.

Chapter 21

Rates and Unit Rates

Definition of Rates

A rate is a ratio that compares two distinct quantities measured in different units by expressing one quantity relative to another. In many cases, the rate is obtained by dividing one measurement by a second, providing a numerical value that describes how one variable changes in relation to the other. Common examples of rates include speed, which relates distance to time, and density, which compares mass to volume. The expression of a rate typically involves two units and forms the basis for understanding proportional relationships in various mathematical and real-world contexts.

Understanding Unit Rates

A unit rate is a specialized form of a rate in which the denominator has a value of one unit. Transforming a general rate into a unit rate simplifies the relationship between the two quantities and allows for a clear interpretation of how much of one quantity corresponds to a single unit of the other. For instance, if a journey covers a distance of 300 miles in 5 hours, the conversion of this rate into a unit rate involves dividing 300 by 5, resulting in 60 miles per hour. This unit rate succinctly describes the distance associated with each individual hour and provides a straightforward basis for comparison across differing scenarios.

Calculating Rates and Unit Rates

The procedure for determining a rate begins with the identification of the two quantities that are being compared. Once these quantities are established, the calculation involves dividing the first quantity (the numerator) by the second quantity (the denominator). In many cases, the resulting fraction represents the rate, and further simplification of that fraction can convert it into a unit rate whenever appropriate. An effective strategy is to first compute the total rate and then adjust the ratio so that the denominator is reduced to one. This systematic approach to calculation is applicable regardless of the units involved, whether they pertain to time, distance, cost, or any other measurable attribute.

Interpreting Rates in Context

Rates and unit rates are used to describe how one quantity changes relative to another in a consistent manner. In various settings, a rate serves as a fundamental tool for comparing different scenarios. The unit rate, in particular, offers a clear and accessible measure by standardizing the comparison to a single unit. By analyzing the rate of speed, for example, an understanding is formed about the distance covered per unit of time. In contexts involving cost or quantity, unit rates provide a method to determine the cost per individual item or the quantity per single unit. These interpretations underscore the concept that the relationship between quantities remains constant even when scaled up or down.

Worked Examples

Consider a scenario where a machine produces 240 widgets in 8 hours. The rate of production is determined by dividing 240 by 8, resulting in a unit rate of 30 widgets per hour. This value conveys that every hour, the machine produces 30 widgets.

In another example, suppose that a store sells 18 notebooks for 27 dollars. To find the cost per notebook, the total cost is divided by the number of notebooks. The calculation, 27 divided by 18, yields a unit rate of 1.5 dollars per notebook. This unit rate clearly identifies the price associated with one notebook.

A further illustration involves the context of travel. If a car travels 150 miles in 3 hours, dividing the total miles by the time taken produces a unit rate of 50 miles per hour. This result indicates that for every hour of travel, the car covers 50 miles, providing a concise measure of its speed.

These examples demonstrate the practical application of rates and unit rates, illustrating both the calculation process and the interpretation of the numerical relationships between different quantities.

Multiple Choice Questions

1. Which of the following best describes a rate?

    (a) A fraction comparing two quantities with identical units.

    (b) A ratio that compares two distinct quantities measured in different units.

    (c) The sum of two measurements.

    (d) A percentage used to calculate discounts.

2. What does the term "unit rate" specifically refer to?

    (a) A rate expressed with a numerator of 1.

    (b) A rate expressed with a denominator of 1.

    (c) The total amount divided by the number of units.

    (d) The product of the two quantities being compared.

3. If a car travels 150 miles in 3 hours, what is its unit rate?

    (a) 40 miles per hour

    (b) 45 miles per hour

    (c) 50 miles per hour

    (d) 55 miles per hour

4. A machine produces 240 widgets in 8 hours. Which of the following represents the machine's production as a unit rate?

    (a) 20 widgets per hour

    (b) 25 widgets per hour

    (c) 30 widgets per hour

    (d) 35 widgets per hour

5. A store sells 18 notebooks for 27 dollars. What is the cost per notebook (unit rate)?

    (a) 1 dollar per notebook
    (b) 1.5 dollars per notebook
    (c) 2 dollars per notebook
    (d) 2.5 dollars per notebook

6. Which of the following is the correct procedure to convert a general rate into a unit rate?

    (a) Multiply the numerator and denominator by the same number.
    (b) Divide both the numerator and the denominator by the denominator.
    (c) Add the numerator to the denominator and then divide by 2.
    (d) Multiply the numerator by the denominator.

7. Why is calculating a unit rate a useful strategy in problem-solving?

    (a) It allows for complex formulas to be ignored.
    (b) It provides a standardized basis for comparing different quantities.
    (c) It eliminates the need for division.
    (d) It converts all measurements to percentages.

Answers:

1. **B: A ratio that compares two distinct quantities measured in different units**
   Explanation: Rates compare two different measurements (like miles per hour or dollars per item) rather than quantities with the same units.

2. **B: A rate expressed with a denominator of 1**
   Explanation: A unit rate standardizes the comparison by showing how much of the numerator corresponds to one unit of the denominator (e.g., 60 miles per hour, where 1 hour is the denominator).

3. **C: 50 miles per hour**
   Explanation: Dividing the total distance (150 miles) by the total time (3 hours) gives the unit rate: 150 ÷ 3 = 50 miles per hour.

4. **C: 30 widgets per hour**
   Explanation: To determine the widgets produced per hour, divide 240 widgets by 8 hours, resulting in 30 widgets per hour.

5. **B: 1.5 dollars per notebook**
   Explanation: The cost per notebook is the total cost divided by the number of notebooks: 27 dollars ÷ 18 = 1.5 dollars per notebook.

6. **B: Divide both the numerator and the denominator by the denominator**
   Explanation: Converting a rate to a unit rate involves dividing both parts of the ratio by the denominator, thus adjusting the denominator to one.

7. **B: It provides a standardized basis for comparing different quantities**
   Explanation: Unit rates simplify comparisons by expressing the relationship per one unit, making it easier to compare situations with different total quantities.

Practice Problems

1. A car travels 300 miles in 5 hours. Find its speed in miles per hour.

2. A printer prints 120 pages in 2 minutes. Determine the printer's speed in pages per minute.

3. A store sells 48 apples for 12 dollars. What is the cost per apple?

4. In a recipe, 4 cups of flour are used to bake 16 cookies. Find the unit rate of flour per cookie (in cups per cookie).

5. A runner completes 8 laps on a 400-meter track in 1 hour and 20 minutes. Calculate her speed in meters per minute.

6. A water pump drains 900 gallons of water in 15 minutes. What is the pump's rate in gallons per minute?

Answers

1. **Solution:** To determine the speed in miles per hour, divide the total distance by the total time:

$$\text{Speed} = \frac{300 \text{ miles}}{5 \text{ hours}} = 60 \text{ miles per hour}.$$

This means that in one hour, the car travels 60 miles.

2. **Solution:** The printer's speed in pages per minute is found by dividing the total number of pages by the time in minutes:

$$\text{Pages per minute} = \frac{120 \text{ pages}}{2 \text{ minutes}} = 60 \text{ pages per minute}.$$

Therefore, the printer prints 60 pages each minute.

3. **Solution:** To find the cost per apple, divide the total cost by the number of apples:

$$\text{Cost per apple} = \frac{12 \text{ dollars}}{48 \text{ apples}} = 0.25 \text{ dollars per apple}.$$

This means that each apple costs 25 cents.

4. **Solution:** The unit rate for flour per cookie is calculated by dividing the total cups of flour by the number of cookies:

$$\text{Flour per cookie} = \frac{4 \text{ cups}}{16 \text{ cookies}} = 0.25 \text{ cups per cookie}.$$

Thus, each cookie requires 0.25 cups of flour.

5. **Solution:** First, convert 1 hour and 20 minutes to minutes:

1 hour and 20 minutes = 60 minutes + 20 minutes = 80 minutes.

Then, compute the total distance for 8 laps:

Total distance = 8 × 400 meters = 3200 meters.

Now, determine the speed in meters per minute:

$$\text{Speed} = \frac{3200 \text{ meters}}{80 \text{ minutes}} = 40 \text{ meters per minute}.$$

This indicates that the runner covers 40 meters in each minute.

6. **Solution:** The water pump's rate is calculated by dividing the total gallons drained by the total time in minutes:

$$\text{Pumping rate} = \frac{900 \text{ gallons}}{15 \text{ minutes}} = 60 \text{ gallons per minute.}$$

Therefore, the pump drains 60 gallons every minute.

Chapter 22

Estimation and Mental Math Strategies

The Concept of Estimation

Estimation is a mathematical technique that involves approximating numerical values in order to simplify calculations and gain insight into the magnitude of quantities. This method does not rely on exact figures; rather, it makes use of rounded values and simplified relationships to indicate approximate results that are often sufficient for checking the reasonableness of an answer or for making rapid decisions in problem solving. Numbers are adjusted to more manageable figures while preserving the underlying proportional relationships, thereby enhancing numerical intuition.

Techniques for Estimation

Several techniques allow for effective estimation during arithmetic operations. These methods take advantage of the properties of numbers and the structure of the base-ten system to make calculations more accessible and efficient.

1 Rounding

Rounding is a fundamental strategy in estimation. The process involves replacing a given number with another number that is close in value but simpler in form. In most cases, numbers are rounded to the nearest ten, hundred, or another convenient unit. For example, in a calculation that involves numbers with many digits, rounding each number to a nearby multiple of ten can streamline the operation. This technique minimizes computational complexity while still leading to an answer that is close to the exact value.

2 Front-End Estimation

Front-end estimation places emphasis on the highest place values. By concentrating on the leftmost digits and making minimal adjustments based on the remaining figures, this method provides a quick approximation of sums or differences. For instance, approximating the addition of two multi-digit numbers by using just the hundreds or tens digits can yield an estimate that guides further detailed calculations. This approach illustrates the effect of significant digits on the overall magnitude of a number.

3 Clustering and Compatible Numbers

When a set of numbers is clustered near a common value, replacing each number with that common value can rapidly yield an estimate. Similarly, the concept of compatible numbers emerges from the observation that certain groups of numbers naturally combine to form round totals. By identifying such groups, mental calculations can be executed with greater speed and confidence. Both of these techniques rely on recognizing patterns in the given values and using those patterns to simplify operations.

Strategies for Mental Calculations

Mental math strategies serve as a bridge between estimation and precise calculation by enabling the manipulation of numbers without recourse to paper or a calculator. These strategies encompass breaking numbers into simpler parts, reorganizing terms, and harnessing the distributive property.

1 Decomposing Numbers

Decomposition involves breaking a complex number into smaller, more manageable parts that are easier to compute. This may include expressing a number as the sum of its tens and ones or separating it into components that simplify multiplication or division.

Through the process of breaking numbers into parts, the mental load is reduced, and operations can be executed piecewise before recombining the results. Each component is handled individually and then aggregated to form the desired answer.

2 Utilizing the Distributive Property

The distributive property enables a large number to be decomposed into factors that simplify multiplication. In mental calculations, this property is used to split a multiplication problem into smaller parts and then add the resulting products. For example, decomposing a product into sums of products of tens and ones allows for a systematic approach to multiplication, where each part contributes to the final estimated result. This method emphasizes both flexibility and structure, rendering complex operations more approachable.

3 Simplifying Complex Operations Mentally

When confronted with complex arithmetic expressions, the use of compatible numbers in addition, subtraction, multiplication, or division can simplify the calculation. By adjusting numbers to establish round figures that are easier to work with, mental calculations become less cumbersome. This strategy involves selecting numbers that are close to the original values yet yield totals that are straightforward to compute. The resultant estimates assist in verifying whether more precise calculations contain plausible values and provide an initial framework for problem solving.

Approximating Results in Arithmetic Operations

Estimation techniques extend across a range of arithmetic operations, providing a consistent method for verifying that detailed computations yield reasonable results.

1 Estimation in Addition and Subtraction

In addition and subtraction, the practice of rounding numbers to nearby multiples of ten or another convenient base allows for quick computation of an approximate result. When several numbers are added together or one number is subtracted from another, retaining the significant digits while ignoring less critical figures provides a mental check that is both accessible and informative. This approximated result serves as a reference point, ensuring that the exact answer falls within an anticipated range.

2 Estimation in Multiplication and Division

Multiplication and division benefit from estimation by reducing the dimensions of the numbers involved. Rounding the factors in a multiplication problem, or the dividend and divisor in a division problem, leads to an estimated outcome that reflects the order of magnitude of the exact answer. This method is particularly useful when the precise multiplication or division would otherwise demand lengthy computation. By operating with rounded values, the underlying proportional relationships remain evident, and the estimated result offers a quick check on the reasonableness of detailed calculations.

Applications in Problem Solving

The strategies for estimation and mental calculations integrate into general problem solving by providing tools that enhance numerical intuition and accelerate computational speed. Whether assessing the impact of individual components in a sum or determining the viability of a subtraction or multiplication operation, these approaches allow for both rapid approximations and momentary checks on the accuracy of complex procedures. The practice of estimation reinforces an intuitive understanding of numbers, underscoring the balance between precision and efficiency in mathematical reasoning.

Multiple Choice Questions

1. What is the primary purpose of estimation in arithmetic?

    (a) To obtain an exact numerical result.

    (b) To quickly approximate values and check the reasonableness of answers.

    (c) To complicate calculations.

(d) To avoid using a calculator.

2. Which technique involves replacing each number with a simpler value—often to the nearest ten or hundred—to simplify calculations?

    (a) Clustering.

    (b) Rounding.

    (c) Front-End Estimation.

    (d) Decomposing Numbers.

3. What does front-end estimation emphasize when approximating a sum or difference?

    (a) The smallest (rightmost) digits.

    (b) The operation symbols.

    (c) The highest (leftmost) place value digits.

    (d) Only the units digit.

4. Which statement best describes the use of clustering and compatible numbers in estimation?

    (a) They involve grouping numbers that automatically round to zero.

    (b) They involve replacing numbers that are close in value with a common approximate value, or selecting numbers that naturally combine to form round totals.

    (c) They involve separating a large number into individual digits.

    (d) They involve ignoring all digits except the highest place value.

5. How does decomposing numbers assist in mental calculations?

    (a) It multiplies numbers by ten to simplify the process.

    (b) It increases the complexity of an arithmetic operation.

    (c) It breaks a complex number into smaller, more manageable parts.

    (d) It rearranges numbers into random groups.

6. How is the distributive property used as a mental math strategy?

   (a) It converts subtraction problems into multiplication problems.
   (b) It separates a multiplication problem into the sum of simpler products.
   (c) It rounds numbers to simplify division.
   (d) It eliminates the need for addition.

7. Why is estimating results in multiplication and division useful?

   (a) It eliminates the need for any precise calculation.
   (b) It provides a quick check to ensure that more detailed computations are reasonable.
   (c) It intentionally generates an inaccurate answer.
   (d) It slows down the calculation process.

Answers:

1. **B: To quickly approximate values and check the reasonableness of answers**
   Estimation is used to simplify calculations and provide a mental check against exact computations, ensuring that answers fall within a reasonable range.

2. **B: Rounding**
   Rounding replaces each number with a nearby, simpler value (such as the nearest ten or hundred) to ease the calculation process and create a quick approximation.

3. **C: The highest (leftmost) place value digits**
   Front-end estimation focuses on the leftmost digits because they contribute most significantly to a number's overall value, providing a good initial approximation.

4. **B: They involve replacing numbers that are close in value with a common approximate value, or selecting numbers that naturally combine to form round totals**
   This technique simplifies calculations by recognizing when numbers are clustered or naturally compatible, making mental arithmetic quicker and easier.

5. **C: It breaks a complex number into smaller, more manageable parts**
   Decomposing numbers reduces the cognitive load by splitting a number into components (such as tens and ones), which can then be manipulated more easily during mental calculations.

6. **B: It separates a multiplication problem into the sum of simpler products**
   By applying the distributive property, a multiplication problem can be broken down into smaller, more manageable parts whose individual products are easier to compute mentally.

7. **B: It provides a quick check to ensure that more detailed computations are reasonable**
   Estimating results in multiplication and division helps verify that the precise calculations are within the expected range, serving as a useful error-checking tool.

Practice Problems

1. Round the number
$$473$$
to the nearest ten.

2. Use front-end estimation to approximate the sum:
$$678 + 425$$
by considering only the largest place values.

3. Estimate the product:

$$48 \times 27$$

by first rounding each factor to the nearest ten and then multiplying.

4. Use compatible numbers to estimate the quotient:

$$142 \div 7$$

by rounding 142 to a nearby number that is easily divisible by 7.

5. Apply the distributive property to mentally compute:

$$36 \times 15$$

by decomposing 36 into two numbers that are easier to multiply with 15.

6. Estimate the result of the subtraction:

$$1984 - 825$$

by rounding each number to a convenient value and then subtracting.

Answers

1. **Problem:** Round 473 to the nearest ten.

   Solution: To round to the nearest ten, examine the ones digit. In 473, the ones digit is 3. Since 3 is less than 5, we round down. This means we keep the tens digit the same and change the ones digit to 0. Therefore,

   $$473 \approx 470.$$

   Explanation: The rounding rule states that when the ones digit is less than 5, the number is rounded down. Thus, 473 becomes 470 when rounded to the nearest ten.

2. **Problem:** Use front-end estimation to approximate the sum 678 + 425.

   Solution: In front-end estimation, focus on the highest place value (the hundreds). For 678, the hundreds digit represents 600, and for 425, it represents 400. Adding these gives:

   $$600 + 400 = 1000.$$

   Explanation: By concentrating on the hundreds place, we simplify the addition and quickly estimate that the sum is approximately 1000. This approach provides a useful check against more detailed calculations.

3. **Problem:** Estimate the product 48 × 27 by rounding each factor to the nearest ten.

   Solution: Round 48 to 50 and 27 to 30. Then multiply:
   $$50 \times 30 = 1500.$$

   Explanation: Rounding both numbers to the nearest ten simplifies the multiplication process. Although the exact product is different, this estimation gives a reasonable idea of the magnitude of the result.

4. **Problem:** Use compatible numbers to estimate the quotient 142 ÷ 7.

   Solution: Notice that 142 is very close to 140, a number that is easily divisible by 7. Dividing:
   $$140 \div 7 = 20.$$

   Explanation: By rounding 142 to 140, we create a compatible number that makes division straightforward. The estimated quotient is 20, which is a good approximation of the actual division.

5. **Problem:** Apply the distributive property to mentally compute 36 × 15 by breaking 36 into two parts.

   Solution: Decompose 36 as 30 + 6. Then apply the distributive property:
   $$36 \times 15 = (30 + 6) \times 15 = 30 \times 15 + 6 \times 15.$$

   Calculate each term:
   $$30 \times 15 = 450 \quad \text{and} \quad 6 \times 15 = 90.$$

   Adding these gives:
   $$450 + 90 = 540.$$

   Explanation: Breaking 36 into 30 and 6 simplifies the multiplication into two easier parts. After computing the products separately and adding them, we obtain the final result of 540.

6. **Problem:** Estimate 1984 825 by rounding each number to a convenient value.

Solution: Round 1984 to 2000 and 825 to 800. Then subtract:
$$2000 - 800 = 1200.$$

Explanation: Rounding the numbers to 2000 and 800 simplifies the subtraction. This estimated result of 1200 provides a quick mental check to ensure that a detailed calculation would yield a similar magnitude.

Chapter 23

Introduction to Variables and Algebraic Expressions

Understanding Variables

Variables serve as symbols that represent numbers in a flexible manner. In arithmetic, operations are performed on concrete numerical values. The introduction of variables transforms these fixed computations into expressions that can accommodate unknown or changeable quantities. Letters such as x, y, or a are commonly utilized to denote these quantities. In this approach, a symbol is embraced as a stand-in for a number, thereby providing the means to express general relationships and patterns without depending on specific values.

1 Definition and Notation

A variable is defined as a placeholder that can assume any numerical value within a given context. In algebraic expressions, variables are typically represented by letters selected from the alphabet. The use of such symbols allows one to write expressions and equations in a compact form. For instance, an expression like 5x indicates that the variable x is being multiplied by 5. The notation solidifies the idea that a variable is not fixed and that its value may be determined later through substitution or given conditions.

2 Role of Variables in Mathematical Generalization

The introduction of variables marks an essential step in the evolution from arithmetic to algebra. This transition allows arithmetic operations to be generalized; rather than computing with specific numerical values, operations can be performed on symbols that represent an entire class of numbers. In this manner, the same algebraic expression can describe multiple situations and is capable of representing a range of numerical relationships. Variables enable the abstraction of mathematical ideas so that the underlying principles become visible regardless of the particular numbers substituted into the expression.

Forming Algebraic Expressions

An algebraic expression is constructed by the combination of numbers, variables, and operation signs. Such expressions serve as precise statements of mathematical ideas and describe relationships between quantities in a condensed format. Algebraic expressions do not merely capture numerical calculations; they encapsulate the structure of the operations themselves and allow for the manipulation of these structures in order to analyze analogies and patterns across different problems.

1 Components of Algebraic Expressions

An algebraic expression is typically composed of several fundamental components. Coefficients are numerical factors that multiply variables; for example, in the term 4y, the number 4 is the coefficient. Constants are fixed numbers that remain unchanged, such as the number 7 in the expression $4y + 7$. Each term in the expression, whether a single variable, a constant, or a product of a number and a variable, contributes to the overall shape of the relationship being modeled. The arrangement of these elements, combined with operation signs such as addition or subtraction, forms a coherent expression that is both compact and robust.

2 Operating with Algebraic Symbols

The manipulation of algebraic expressions follows systematic arithmetic operations. Variables, despite representing unknown quantities, adhere to the general mathematical rules governing addition, subtraction, multiplication, and division. When expressions include like terms—terms that contain the same variable raised to the same power—their coefficients can be combined to simplify the expression. Through the artful use of algebraic operations, expressions can be rearranged, expanded, and simplified while preserving their intrinsic relationships. This systematic treatment of symbols paves the way for solving equations and exploring mathematical properties in a broad and generalized context.

Transitioning from Arithmetic to Algebra

The evolution from arithmetic to algebra is marked by the generalization of numerical operations through the introduction of variables. In arithmetic, computations are confined to known numbers, resulting in fixed outcomes. The use of variables in algebra permits the formulation of expressions that stand as representations of relationships regardless of the specific values involved. This shift creates a framework wherein operations can be performed symbolically, enabling the exploration of patterns, the derivation of formulas, and the analysis of relationships in a more abstract, yet precise, manner.

1 Generalization of Arithmetic Operations

With the inclusion of variables, familiar arithmetic operations transcend their original limitations and become components of broader algebraic expressions. An operation such as addition, when reinterpreted in an algebraic context, may involve one or more variables alongside constants. For instance, the sum expressed as $x + 3$ embodies a flexible arithmetic operation where the unknown quantity x is seamlessly integrated with a fixed number. The abstraction achieved by such generalizations demonstrates the power of algebra to represent a wide array of quantitative scenarios with a single, adaptable format.

2 Mathematical Modeling Using Variables

Algebraic expressions offer a versatile method for modeling diverse mathematical relationships. Variables act as dynamic placeholders that capture the essence of changing quantities, while the structure of the expression denotes how these quantities interact. This capacity to model relationships is integral to understanding the behavior of numerical systems in both theoretical constructs and practical applications. By representing relationships symbolically, the expressions provide a foundation for analyzing scenarios where direct computation is impractical or where general rules must be established prior to the introduction of specific numerical values.

Multiple Choice Questions

1. What is a variable in algebra?

    (a) A fixed number that never changes.

    (b) A symbol that represents an unknown or changeable number.

    (c) An arithmetic operation like addition or multiplication.

    (d) A type of geometric shape.

2. Which of the following best illustrates an algebraic expression?

    (a) 5x + 3

    (b) 7 + 2

    (c) x - y = 4

    (d) 9

3. In the term 4y, what does the number 4 represent?

    (a) The variable.

    (b) The coefficient.

    (c) The constant.

    (d) The exponent.

4. What is the result of combining like terms in the expression 2x + 3x?

    (a) 5x

    (b) 6x

    (c) 5

    (d) 6

5. What does it mean to generalize arithmetic operations by using variables?

    (a) Replacing fixed numbers with symbols to represent a range of values.

    (b) Limiting computations only to addition and subtraction.

    (c) Ensuring every operation results in a numerical answer immediately.

    (d) Making arithmetic more difficult without any benefits.

6. Which of the following are fundamental components of an algebraic expression?

    (a) Variables, coefficients, constants, and operation signs.

    (b) Only numbers and operation signs.

    (c) Only variables and operation signs.

    (d) Only coefficients and constants.

7. How do variables assist in mathematical modeling?

    (a) They simplify computations by providing exact numbers.

    (b) They allow one to express relationships in a generalized form.

    (c) They ensure that every model produces a fixed outcome.

    (d) They eliminate the need for forming equations.

Answers:

1. **B: A symbol that represents an unknown or changeable number.**
   Explanation: Variables act as placeholders that can take on different numerical values, allowing us to create general expressions and solve problems even when exact numbers aren't specified.

2. **A: $5x + 3$.**
   Explanation: The expression $5x + 3$ is algebraic because it includes a variable (x) along with a coefficient (5), a constant (3), and an operation (addition). The other choices either lack a variable or represent an equation.

3. **B: The coefficient.**
   Explanation: In the term 4y, the number 4 multiplies the variable y. This multiplier is known as the coefficient.

4. **A: 5x.**
   Explanation: When combining like terms, you add their coefficients. Here, 2x + 3x results in (2 + 3)x, which simplifies to 5x.

5. **A: Replacing fixed numbers with symbols to represent a range of values.**
   Explanation: Using variables allows us to generalize arithmetic operations by substituting numbers with symbols. This makes it possible to represent many scenarios with one expression.

6. **A: Variables, coefficients, constants, and operation signs.**
   Explanation: An algebraic expression is built from these key elements. Variables represent unknowns, coefficients multiply the variables, constants provide fixed values, and operation signs dictate the arithmetic processes.

7. **B: They allow one to express relationships in a generalized form.**
   Explanation: Variables let us model real-world and mathematical relationships without committing to specific numerical values. This generalization is key in forming equations and exploring patterns.

Practice Problems

1. Identify the variable(s), coefficient(s), and constant in the algebraic expression:
$$3a + 8.$$

2. Write an algebraic expression for the phrase "five more than twice a number x."

3. Simplify the algebraic expression by combining like terms:
$$2x + 3x - 4 + 7.$$

4. Evaluate the expression
$$4y + 2$$
when the variable takes the value $y = 3$. (Remember to substitute and simplify.)

5. Translate the sentence "The product of a number n and 6, decreased by 4" into an algebraic expression.

6. Express the relationship "The perimeter of a rectangle is twice the sum of its length l and width w" using variables.

Answers

1. For the expression
$$3a + 8,$$
   Solution: The term $3a$ includes a coefficient and a variable. Here, the number 3 is the coefficient and a is the variable. The number 8 is a constant because it is not attached to any variable. Therefore, the variable is a, the coefficient is 3, and the constant is 8.

2. For the phrase "five more than twice a number x": **Solution:** The phrase "twice a number x" means $2x$. When we say "five more than" that quantity, we add 5 to $2x$. Thus, the algebraic expression is:
$$2x + 5.$$
   This expression shows that 5 is added to two times the number x.

3. For the expression
$$2x + 3x - 4 + 7,$$
   Solution: First, combine the like terms. The terms $2x$ and $3x$ are like terms, and their sum is:
$$2x + 3x = 5x.$$
   Next, combine the constant terms -4 and 7:
$$-4 + 7 = 3.$$
   This yields the simplified expression:
$$5x + 3.$$

4. For evaluating the expression
$$4y + 2$$
   when $y = 3$: **Solution:** Substitute 3 for y in the expression:
$$4(3) + 2.$$
   Multiply 4 by 3:
$$12 + 2.$$
   Finally, add 12 and 2 to obtain:
$$14.$$
   Therefore, the value of the expression when $y = 3$ is 14.

5. For the sentence "The product of a number n and 6, decreased by 4": **Solution:** The phrase "the product of a number n and 6" translates to $6n$. "Decreased by 4" means that 4 is subtracted from that product. Hence, the algebraic expression is:
$$6n - 4.$$
   This expression accurately represents the given sentence.

6. For the relationship "The perimeter of a rectangle is twice the sum of its length l and width w": **Solution:** The sum of the length l and width w is expressed as $l + w$. Since the perimeter is twice this sum, we multiply the entire sum by 2. The algebraic expression for the perimeter P is:
$$P = 2(l + w).$$
   This expression correctly models the relationship between a rectangle's dimensions and its perimeter.

Chapter 24

Writing and Evaluating Algebraic Expressions

Translating Verbal Descriptions into Algebraic Expressions

1 Identifying Keywords and Mathematical Operations

Verbal descriptions of mathematical situations contain specific keywords that indicate various operations and relationships between quantities. Terms such as "sum," "difference," "product," and "quotient" correspond to addition, subtraction, multiplication, and division respectively. Phrases like "increased by," "decreased by," and "more than" signal the addition or subtraction of numbers, while expressions such as "times" or "of" denote multiplication. The systematic interpretation of these key words results in an accurate conversion of a word problem into an algebraic formulation.

2 Assigning Variables and Forming Expressions

The process begins by designating a variable, typically represented by a letter such as x, y, or n. This variable serves as a placeholder for an unknown or changeable value mentioned in the verbal description. For instance, the phrase "twice a number" is translated into the term 2x if x is chosen to represent the number. In cases where the description indicates an addition or subtraction of a quantity, the constant is combined with the variable term through the appropriate operation. The careful ordering of identified components results in a coherent expression that mirrors the original verbal statement.

3 Detailed Process for Translation

After identifying all keywords and assigning the appropriate variables, the translation proceeds through the sequential combination of these elements. The mathematical operations are written using standard algebraic notation. For each segment of the verbal description, the corresponding mathematical operation is denoted using symbols such as $+$, $-$, or $\times$, ensuring that the natural order of operations is maintained. For example, the verbal phrase "five more than three times a number" is systematically transformed into the expression $3x + 5$. This process requires careful parsing of the language so that all numerical and operational parts are accurately captured in the resultant algebraic expression.

Evaluating Algebraic Expressions for Given Values

1 Method of Substitution and Arithmetic Computation

Evaluating an algebraic expression involves replacing the variable with a specific numerical value and performing the indicated arithmetic operations. This process, known as substitution, adheres to the established order of operations. Given the expression $3x + 4$ and a prescribed value for x, the first step requires substituting the value into the expression. The ensuing arithmetic is then completed by calculating the products and sums in accordance with conventional mathematical rules. For example, substituting 5 for x yields $3(5) + 4$, resulting in an evaluation of 19.

2 Worked Examples and Step-by-Step Evaluation

Consider the expression $2x - 7$, where the value of x is provided as 6. Substitution leads to $2(6) - 7$. The multiplication is performed first, resulting in 12, followed by the subtraction to produce the final result of 5. In another example, an expression such as $4y + 3$ can be evaluated by substituting y with a given number, say 2, to obtain $4(2) + 3$. The product 8, when added to 3, gives a result of 11. Each example illustrates the systematic replacement of the variable with its prescribed value, followed by a clear application of arithmetic operations to arrive at a simplified numerical outcome.

3 Ensuring Accuracy through Verification

Accurate evaluation of algebraic expressions demands meticulous adherence to the order of operations. It is essential that multiplications and divisions are performed before additions and subtractions. Verification of each step provides a methodical approach to ensure that the final result reflects the intended translation and calculation. When multiple terms and operations are present, each substitution and arithmetic step is double-checked. This methodical process assists in avoiding common pitfalls, such as misinterpretation of the verbal description or error in carrying out arithmetic operations, thereby ensuring a reliable and precise evaluation.

Multiple Choice Questions

1. Which algebraic expression best represents the verbal description "five more than twice a number"?

    (a) $2x + 5$

    (b) $5x + 2$

    (c) $2(x + 5)$

    (d) $x + 5$

2. Which keyword in a verbal description most clearly indicates multiplication?

    (a) Sum

    (b) Difference

    (c) Times

    (d) More than

3. What is the first step in evaluating an algebraic expression using substitution?

    (a) Add all the terms together.

(b) Substitute the given numerical values for the variables.

(c) Multiply the coefficients in the expression.

(d) Ignore the order of operations.

4. Why is assigning a variable to an unknown quantity important when forming algebraic expressions?

   (a) It automatically simplifies the arithmetic.

   (b) It provides a clear symbolic representation for an unknown value.

   (c) It adds unnecessary complexity to a problem.

   (d) It ensures that all terms have numerical coefficients.

5. Translate the verbal statement "seven less than four times a number" into an algebraic expression.

   (a) 7 - 4x

   (b) 4x - 7

   (c) 4 - 7x

   (d) 7 ÷ 4x

6. Evaluate the expression 4y + 2 when y = 3.

   (a) 10

   (b) 12

   (c) 14

   (d) 16

7. Why is it important to follow the order of operations when evaluating an algebraic expression?

   (a) It allows you to perform operations in any sequence.

   (b) It is only necessary for very complex expressions.

   (c) It ensures a consistent and correct evaluation.

   (d) It simplifies the expression automatically.

Answers:

1. **A: 2x + 5** "Twice a number" is represented by 2x, and "five more than" tells us to add 5. Thus, the expression is 2x + 5.

2. **C: Times** The word "times" directly indicates multiplication in a verbal description, guiding the translation into an algebraic expression.

3. **B: Substitute the given numerical values for the variables** Evaluating an expression requires that you first replace each variable with its prescribed number before performing any arithmetic operations.

4. **B: It provides a clear symbolic representation for an unknown value** Assigning a variable (such as x or y) is fundamental because it allows you to denote an unknown quantity and form expressions that model the problem.

5. **B: 4x - 7** "Four times a number" converts to 4x, and the phrase "seven less than" indicates that 7 is to be subtracted from 4x, resulting in 4x - 7.

6. **C: 14** By substituting $y = 3$ into the expression $4y + 2$, you calculate $4(3) + 2 = 12 + 2 = 14$.

7. **C: It ensures a consistent and correct evaluation** Following the order of operations (substitution, then performing multiplication/division before addition/subtraction) is crucial to arrive at the correct answer every time.

Practice Problems

1. Translate the verbal description "Seven less than twice a number" into an algebraic expression.

2. Translate the verbal description "Three times the sum of a number and four" into an algebraic expression.

3. Translate the verbal description "The quotient of the product of five and a number, and the number decreased by three" into an algebraic expression.

4. Evaluate the algebraic expression
$$3x + 4$$
for x = 5.

5. Evaluate the algebraic expression
$$2(3y - 4) + 5$$
for y = 3.

6. Translate the verbal description "Four more than twice a number" into an algebraic expression and evaluate it for the number 8.

Answers

1. **Solution:** Let x represent the number. The phrase "twice a number" translates to 2x, and "seven less than" means we subtract 7. Thus, the algebraic expression is:

$$2x - 7.$$

   Explanation: The operation order follows the verbal description: first, compute twice the number (2x) and then subtract 7 to obtain 2x - 7.

2. **Solution:** Let x denote the number. The phrase "the sum of a number and four" is written as (x + 4), and "three times" this sum is expressed as:

$$3(x + 4).$$

   Explanation: Grouping the sum (x + 4) within parentheses ensures that the multiplier 3 is applied to both x and 4, accurately reflecting the verbal description.

3. **Solution:** Let x be the number. "The product of five and a number" is 5x, and "the number decreased by three" is (x - 3). The term "quotient" indicates division, so the expression becomes:

$$\frac{5x}{x - 3}.$$

   Explanation: Here, 5x is divided by (x - 3) to directly represent the stated relationship. Parentheses in the denominator clarify that the subtraction applies to x before division.

4. **Solution:** To evaluate the expression
$$3x + 4$$
for x = 5, substitute 5 in place of x:
$$3(5) + 4 = 15 + 4 = 19.$$

Explanation: According to the order of operations, multiply 3 by 5 to get 15 and then add 4 to reach the final value of 19.

5. **Solution:** Substitute y = 3 into the expression
$$2(3y - 4) + 5:$$
$$2(3(3) - 4) + 5 = 2(9 - 4) + 5 = 2(5) + 5 = 10 + 5 = 15.$$

Explanation: First, replace y with 3 to compute inside the parentheses: 3(3) equals 9, then subtract 4 to get 5. Multiply 5 by 2 and finally add 5 to arrive at 15.

6. **Solution:** Let x represent the number. "Twice a number" means 2x, and "four more than" instructs us to add 4:
$$2x + 4.$$
Now, evaluate this expression for x = 8:
$$2(8) + 4 = 16 + 4 = 20.$$

Explanation: The translation yields 2x + 4. Substituting 8 for x gives 2(8) = 16, and adding 4 results in 20.

Chapter 25

The Distributive Property

Definition and Fundamental Concepts

1 Definition of the Distributive Property

The distributive property is a central principle in algebra that connects multiplication with addition and subtraction. In expressions where a single factor multiplies a sum or difference, the multiplication is applied to each term separately. This property is formally expressed as
$$a(b + c) = ab + ac,$$
and in the case of subtraction as
$$a(b - c) = ab - ac.$$
This rule permits the elimination of parentheses by ensuring that the multiplier is consistently applied to every term within the grouped expression.

2 Underlying Rationale

The foundation of the distributive property lies in the idea that multiplication represents repeated addition. When a number multiplies a sum, it is equivalent to adding multiple copies of each term scaled by that number. This consistency in scaling creates an equivalence between the factored and expanded forms of an expression. Such logical coherence is essential in algebra, providing a clear method for both expanding expressions and later simplifying them through the combination of like terms.

Expanding Algebraic Expressions

1 Procedure for Expansion

Expanding an algebraic expression involves applying the distributive property to remove parentheses and rewrite the expression as a sum or difference of individual terms. For example, consider

$$3(x+5).$$

Application of the distributive property yields

$$3 \cdot x + 3 \cdot 5 = 3x + 15.$$

Likewise, in an expression where subtraction occurs,

$$-2(y-4)$$

expansion leads to

$$-2 \cdot y + (-2) \cdot (-4) = -2y + 8.$$

Each term inside the parentheses is multiplied by the external factor, ensuring that the expanded form maintains the same value as the original expression.

2 Expansion with Multiple Terms

When the expression inside the parentheses contains more than two terms, the distributive property is applied uniformly to each term. For instance,
$$4(a + 2b - 3c)$$
is expanded by distributing the factor 4:

$$4a + 8b - 12c.$$

This approach guarantees that every term is accounted for, thereby providing a clear path from the factored form to the expanded form.

Simplification Through the Distributive Property

1 Combining Like Terms After Expansion

After expansion, the process of simplification generally involves combining like terms. For example, starting from
$$2(3x + 4) + 5x,$$
expansion gives
$$6x + 8 + 5x.$$
Combining the terms with the variable x results in
$$(6x + 5x) + 8 = 11x + 8.$$

This procedure not only simplifies the expression but also prepares it for further algebraic manipulation.

2 Reverse Application: Factoring

The distributive property also functions in reverse to factor common terms from an expression. Given
$$6x + 9,$$
a common factor of 3 can be factored out to yield
$$3(2x + 3).$$

This reverse process, often referred to as factoring, exemplifies the dual nature of the distributive property. It demonstrates how an expanded expression can be condensed back into a more compact, factored form, reinforcing the connection between these algebraic processes.

3 Detailed Example

Consider the expression
$$5(2x - 3) - 2(x + 4).$$
The first step is to apply the distributive property to each grouped term:
$$5 \cdot 2x - 5 \cdot 3 - 2 \cdot x - 2 \cdot 4 = 10x - 15 - 2x - 8.$$

Next, the like terms are combined:
$$(10x - 2x) + (-15 - 8) = 8x - 23.$$

This example illustrates the systematic procedure of first expanding the expression using the distributive property and then simplifying the result by merging like terms, thereby yielding a clear and concise algebraic expression.

Multiple Choice Questions

1. Which property is represented by the equation a(b + c) = ab + ac?

    (a) Identity Property
    (b) Commutative Property
    (c) Distributive Property
    (d) Associative Property

2. What is the expanded form of the expression 3(x - 4)?

    (a) 3x - 4
    (b) 3x - 12
    (c) 3x + 4
    (d) 3x + 12

3. How does the distributive property help simplify algebraic expressions?

    (a) It rearranges the order of operations.
    (b) It removes parentheses by multiplying each term and then allows like terms to be combined.
    (c) It transforms subtraction into addition.
    (d) It replaces factors with exponents.

4. In the expression -2(y - 4), why does multiplying -2 and -4 yield a positive product?

    (a) Because subtracting a negative number equals adding a positive.
    (b) Because multiplying two negative numbers always gives a positive result.

- (c) Because the negative sign outside the parentheses cancels the negatives inside.
- (d) Because order of operations forces the negatives to vanish.

5. Which process demonstrates the reverse use of the distributive property (factoring)?
   - (a) Converting 4(a + 2b - 3c) into 4a + 8b - 12c.
   - (b) Transforming 6x + 9 into 3(2x + 3).
   - (c) Expanding -2(y - 4) into -2y + 8.
   - (d) Combining 2(3x + 4) with 5x.

6. When applying the distributive property to the expression 4(a + 2b - 3c), what action is performed on each term inside the parentheses?
   - (a) Each term is divided by 4.
   - (b) Each term is multiplied by 4.
   - (c) Each term has 4 added to it.
   - (d) Each term has 4 subtracted from it.

7. What is the fully simplified form of the expression 5(2x - 3) - 2(x + 4)?
   - (a) 8x - 17
   - (b) 8x - 23
   - (c) 10x - 23
   - (d) 10x - 17

Answers:

1. **C: Distributive Property** This property shows that multiplying a number by a sum is equivalent to multiplying the number by each addend and then adding the results.

2. **B: 3x - 12** Expanding 3(x - 4): Multiply 3 by x to get 3x, and 3 by -4 to obtain -12.

3. **B: It removes parentheses by multiplying each term and then allows like terms to be combined** The distributive property enables us to eliminate parentheses by distributing the multiplier across each term inside, simplifying further by combining like terms when possible.

4. **B: Because multiplying two negative numbers always gives a positive result** In -2(y - 4), when -2 is multiplied by -4, the product is positive 8 since the product of two negatives is positive.

5. **B: Transforming 6x + 9 into 3(2x + 3)** This is an example of factoring, which is the reverse of distributing. A common factor (3) is taken out from both terms.

6. **B: Each term is multiplied by 4** The distributive property requires multiplying the external factor (4) by every term inside the parentheses (a, 2b, and -3c).

7. **B: 8x - 23** First, expand the expression: 5(2x - 3) gives 10x - 15 and -2(x + 4) gives -2x - 8. Combining like terms: (10x - 2x) = 8x and (-15 - 8) = -23.

Practice Problems

1. Expand the expression:

$$5(x+8)$$

2. Expand the expression:

$$-3(2y-7)$$

3. Expand the expression:
$$4(a + 2b - 3c)$$

4. Factor the expression using the reverse application of the distributive property:
$$9m + 27$$

5. Simplify the expression:
$$2(3x + 4) + 5x$$

6. Simplify the expression:
$$4(2x-3) - 3(2x-3)$$

Answers

1. **Problem:** Expand
$$5(x+8)$$

   Solution: Apply the distributive property by multiplying 5 by each term inside the parentheses:
$$5 \cdot x + 5 \cdot 8 = 5x + 40.$$

   Explanation: The distributive property states that multiplying a number by a sum means multiplying the number with each addend separately and then adding the results.

2. **Problem:** Expand
$$-3(2y-7)$$

   Solution: Distribute -3 to both terms inside the parentheses:
$$-3 \cdot 2y + (-3) \cdot (-7) = -6y + 21.$$

   Explanation: Notice that multiplying a negative by a negative gives a positive result. Here, -3 times 2y gives -6y, and -3 times -7 gives +21.

3. **Problem:** Expand
$$4(a+2b-3c)$$

Solution: Multiply 4 by each term inside the parentheses:

$$4 \cdot a + 4 \cdot 2b + 4 \cdot (-3c) = 4a + 8b - 12c.$$

Explanation: Each term inside the parentheses is multiplied by 4, ensuring that the operation affects all parts of the expression uniformly.

4. **Problem:** Factor

$$9m + 27$$

using the reverse application of the distributive property.

Solution: Identify the common factor in both terms. Both 9m and 27 are divisible by 9:

$$9m + 27 = 9(m) + 9(3) = 9(m + 3).$$

Explanation: Factoring is the reverse process of distribution. By extracting the common factor 9, we condense the expression into a product of 9 and the binomial (m + 3).

5. **Problem:** Simplify

$$2(3x + 4) + 5x$$

Solution: First, distribute the 2:

$$2 \cdot 3x + 2 \cdot 4 = 6x + 8.$$

Then, add the like term 5x:

$$6x + 8 + 5x = (6x + 5x) + 8 = 11x + 8.$$

Explanation: After expanding the parentheses by distributing 2, combine the like terms (6x and 5x) to arrive at the simplified expression.

6. **Problem:** Simplify

$$4(2x - 3) - 3(2x - 3)$$

Solution: Notice that both terms share the common factor (2x - 3). Factor (2x - 3) out:

$$4(2x - 3) - 3(2x - 3) = (4 - 3)(2x - 3) = 1 \cdot (2x - 3) = 2x - 3.$$

Alternatively, you could expand each term and then combine like terms:

$4(2x)-4(3)-3(2x)+3(3) = 8x-12-6x+9 = (8x-6x)+(-12+9) = 2x-3.$

Explanation: Recognizing the common factor (2x - 3) allows you to factor it out directly. This method shows the efficiency of the reverse application of the distributive property and simplifies the expression quickly.

Chapter 26

Combining Like Terms

Definition of Like Terms

An algebraic term consists of a numerical coefficient and a variable component. When multiple terms possess identical variable components raised to the same power, these terms are defined as like terms. The numerical coefficients in these terms may differ, but the variable portion remains consistent. This concept forms an essential foundation in algebraic manipulation by enabling expressions to be simplified through the combination of these similar elements.

Identifying Like Terms in Algebraic Expressions

In any algebraic expression, several terms are often present, and determining which among these are like terms involves a close examination of their variable portions. For instance, the terms 7x, -3x, and 5x are like terms because each involves the same variable x raised to the first power. In contrast, terms such as 4x and $4x^2$ are not like terms; the presence of a different exponent in $4x^2$ indicates that its variable component does not match that of 4x. Such careful observation aids in clearly distinguishing the terms that can be combined.

Process of Combining Like Terms

Once the like terms within an expression have been identified, the next step is to combine them by performing arithmetic operations on their coefficients. The variable component remains unaltered during this process. For example, in the expression

$$3x + 5 + 2x - 8,$$

the terms 3x and 2x are like terms and are combined through the addition of their coefficients, resulting in

$$(3 + 2)x = 5x.$$

Similarly, the constant terms 5 and -8 combine to yield -3 through simple arithmetic addition. The expression thus simplifies to

$$5x - 3.$$

This systematic approach ensures a clear reduction of the expression, facilitating further algebraic procedures.

Worked Examples

Consider the expression

$$4y + 3 + 7y - 5 + 2y.$$

The like terms containing the variable y are 4y, 7y, and 2y. Adding their coefficients produces

$$4 + 7 + 2 = 13,$$

resulting in the combined term

$$13y.$$

The constant terms 3 and -5 combine to form -2. Therefore, the simplified expression is
$$13y - 2.$$

Another example is the expression

$$2x + 6 - 5x + 3x - 4.$$

The like terms having the variable x—namely 2x, -5x, and 3x—combine as follows:
$$(2 - 5 + 3)x = 0x,$$
which effectively cancels out the variable terms. The constant terms 6 and -4 add up to 2, leaving the simplified expression as

$$2.$$

These examples serve to illustrate how combining like terms can transform a complex algebraic expression into a more manageable form.

Common Pitfalls and Considerations

Errors in combining like terms often arise from overlooking the requirement that the variable components must be exactly identical, including the exponent. For example, misconstruing 2x and $2x^2$ as like terms may lead to incorrect simplifications. Additionally, mismanaging arithmetic operations—especially the handling of negative numbers—can result in computational errors. A careful, methodical approach is advisable: first, identify and clearly group the like terms; next, perform the necessary arithmetic on the coefficients; and finally, rewrite the expression with the combined terms. Attention to these details ensures that simplification is both accurate and effective.

Multiple Choice Questions

1. Which of the following best defines "like terms" in an algebraic expression?

    (a) Terms that always have the same numerical coefficient.

    (b) Terms that have identical variable parts, including the same exponents.

    (c) Terms that appear next to each other in the expression regardless of their variables.

    (d) Terms that sum to zero when combined.

2. Which pair of terms is considered like terms?

    (a) 3x and $3x^2$

(b) 4y and -2y

(c) 7z and 7

(d) 5a and 5b

3. When simplifying the expression

$$3x + 5 + 2x - 8,$$

which of the following is the correct result?

(a) 5x - 3

(b) 5x + 13

(c) 5x - 13

(d) 5x + 3

4. Consider the expression

$$4a + 3 - 2a + 5.$$

What is the correct process to simplify this expression?

(a) Combine 4a and -2a to obtain 2a; add 3 and 5 to get 8; the expression simplifies to 2a + 8.

(b) Multiply 4a by -2a and then add the constant terms.

(c) Add all numerical values (including those with variables) together to obtain 10a.

(d) Combine 4a with 3 to obtain 7a, leaving the remaining terms unchanged.

5. In the process of combining like terms, which component must remain unchanged?

(a) The numerical coefficient

(b) The sign of each term

(c) The variable part along with its exponent

(d) The order of the terms in the expression

6. Which of the following represents an incorrect method when combining like terms?

(a) Adding the numerical coefficients while keeping the variable part unchanged

(b) Combining constant terms separately

(c) Adding the exponents of like terms together when combining them

(d) Grouping like terms together before performing arithmetic operations

7. Simplify the expression
$$2x + 8 - 3x + 2 - 5.$$

What is the correct simplified form?

(a) -x + 5
(b) -x + 15
(c) x + 5
(d) 5x + 5

Answers:

1. **B: Terms that have identical variable parts, including the same exponents**
   Explanation: Like terms are defined as having the same variable component with exactly the same exponent, even though their numerical coefficients may differ.

2. **B: 4y and -2y**
   Explanation: Both terms contain the variable y raised to the first power. The other pairs either have different exponents or different variables.

3. **A: 5x - 3**
   Explanation: First, combine the x terms: 3x + 2x = 5x, then combine the constant terms: 5 - 8 = -3. The final simplified expression is 5x - 3.

4. **A: Combine 4a and -2a to obtain 2a; add 3 and 5 to get 8; the expression simplifies to 2a + 8.**
   Explanation: Group the like terms by combining the terms with the variable a (4a - 2a = 2a) and the constants (3 + 5 = 8), which results in 2a + 8.

5. **C: The variable part along with its exponent**
   Explanation: When combining like terms, only the numerical coefficients are combined. The variable part (including its exponent) must remain unchanged.

6. **C: Adding the exponents of like terms together when combining them**
   Explanation: The correct method is to add or subtract the coefficients of like terms, not to add the exponents. The variable part is left unchanged during the process.

7. **A: -x + 5**
   Explanation: Combine the x terms: 2x - 3x = -x; then combine the constants: 8 + 2 - 5 = 5. The simplified expression is -x + 5.

Practice Problems

1. Simplify the expression:
$$3x + 7 - 2x + 4.$$

2. Simplify the expression:
$$5y - 2 + 3 - 4y + 8y.$$

3. Simplify the expression:
$$7a + 3b - 2a + 4 - 5b + 6.$$

4. Simplify the expression:
$$2x^2 + 3x - 4x^2 + 5 - 7x + 2.$$

5. Simplify the expression:
$$9 - 3y + 2y^2 - 4 + y - 2y^2.$$

6. Simplify the expression:
$$4m + 6n - 2m + 3n + 8 - 10.$$

Answers

1. **Problem:** Simplify
$$3x + 7 - 2x + 4.$$

   Solution:
   First, group the like terms. The terms with the variable x are:
   $$3x - 2x,$$
   which combine to give:
   $$(3 - 2)x = x.$$
   Next, combine the constant terms:
   $$7 + 4 = 11.$$
   Therefore, the simplified expression is:
   $$x + 11.$$

2. **Problem:** Simplify
$$5y - 2 + 3 - 4y + 8y.$$

   Solution:
   Begin by grouping the like terms with the variable y:
   $$5y - 4y + 8y.$$

Adding the coefficients:
$$(5 - 4 + 8)y = 9y.$$

Next, group the constants:
$$-2 + 3 = 1.$$

Thus, the expression simplifies to:
$$9y + 1.$$

3. **Problem:** Simplify
$$7a + 3b - 2a + 4 - 5b + 6.$$

   Solution:
   First, combine the like terms for each variable. For the a terms:
   $$7a - 2a = 5a.$$
   For the b terms:
   $$3b - 5b = -2b.$$
   Then, combine the constant terms:
   $$4 + 6 = 10.$$
   The simplified expression is:
   $$5a - 2b + 10.$$

4. **Problem:** Simplify
$$2x^2 + 3x - 4x^2 + 5 - 7x + 2.$$

   Solution:
   Group the like terms by their variable components. For the x^2 terms:
   $$2x^2 - 4x^2 = -2x^2.$$
   For the x terms:
   $$3x - 7x = -4x.$$
   Then, add the constant terms:
   $$5 + 2 = 7.$$
   Hence, the simplified expression is:
   $$-2x^2 - 4x + 7.$$

5. **Problem:** Simplify
$$9 - 3y + 2y^2 - 4 + y - 2y^2.$$

   Solution:
   Identify and combine the like terms. First, notice that the y^2 terms cancel each other:
$$2y^2 - 2y^2 = 0.$$

   Next, combine the y terms:
$$-3y + y = -2y.$$

   Then, combine the constant terms:
$$9 - 4 = 5.$$

   The final simplified expression is:
$$-2y + 5.$$

6. **Problem:** Simplify
$$4m + 6n - 2m + 3n + 8 - 10.$$

   Solution:
   Group the terms by their variables and constants. For the m terms:
$$4m - 2m = 2m.$$

   For the n terms:
$$6n + 3n = 9n.$$

   Finally, combine the constant terms:
$$8 - 10 = -2.$$

   Thus, the simplified expression is:
$$2m + 9n - 2.$$

Chapter 27

Solving One-Step Equations

Definition and Characteristics of One-Step Equations

A one-step equation is an algebraic equation in which the variable is isolated by means of a single arithmetic operation. Such equations typically contain only one operation that must be undone in order to determine the value of the unknown. Typical forms include equations where a constant is added to or subtracted from a variable (for example, $x+5=12$ or $x-3=8$), as well as equations where the variable is multiplied by or divided by a constant (such as $4x=20$ or $\frac{x}{3}=7$). The process of solving these equations depends on the proper application of an inverse operation that cancels the existing operation affecting the variable.

Inverse Operations and Their Application

The fundamental technique in solving a one-step equation involves the application of an inverse operation. Inverse operations are operations that reverse the effect of another. When a constant is added to a variable, subtraction is applied as the inverse operation, and similarly, when a constant is subtracted, addition is used. For equations involving multiplication, division acts as the inverse, and for those involving division, multiplication is employed. The application of an inverse operation must be done equally to both sides of the equation in order to retain the balance of the equality. For example, given an equation of the form

$$x + c = d,$$

subtracting c from both sides leads to

$$x = d - c.$$

This method ensures that the original relationship between the expressions remains valid.

Solving Equations Involving Addition and Subtraction

Equations that incorporate either addition or subtraction with the variable require the performance of the corresponding inverse operation. Consider the equation

$$x + 7 = 15.$$

The operation of addition is reversed by subtracting 7 from both sides, which yields

$$x = 15 - 7.$$

Similarly, for an equation given by

$$x - 4 = 9,$$

addition of 4 to both sides results in

$$x = 9 + 4.$$

In each example, the arithmetic performed on both sides of the equation successfully isolates the variable.

Solving Equations Involving Multiplication and Division

In one-step equations where the variable is multiplied or divided by a constant, the inverse operations of division and multiplication, respectively, are employed. Consider an equation of the form

$$5x = 30.$$

Here, division by 5 is the inverse operation needed, and dividing both sides by 5 yields
$$x = \frac{30}{5}.$$
In an equation such as
$$\frac{x}{8} = 3,$$
multiplication by 8 is applied on both sides to obtain
$$x = 3 \times 8.$$

These operations effectively remove the constant factor attached to the variable, thereby isolating the unknown.

Worked Examples

Examine the following examples which demonstrate the systematic approach to solving one-step equations:

Example 1: Addition Operation

Given the equation
$$x + 6 = 14,$$
the inverse operation is subtraction. Subtracting 6 from both sides results in
$$x = 14 - 6,$$
thus
$$x = 8.$$

Example 2: Subtraction Operation

For the equation
$$x - 5 = 10,$$
the inverse operation is addition. Adding 5 to both sides leads to
$$x = 10 + 5,$$
so that
$$x = 15.$$

Example 3: Multiplication Operation

Consider the equation
$$3x = 21.$$
Division by 3 on both sides gives
$$x = \frac{21}{3},$$
which simplifies to
$$x = 7.$$

Example 4: Division Operation

In the equation
$$\frac{x}{4} = 5,$$
the variable is isolated by multiplying both sides by 4. This yields
$$x = 5 \times 4,$$
and finally
$$x = 20.$$

Observations and Common Error Patterns

When performing the inverse operation, it is essential that the same arithmetic operation is applied to each side of the equation to maintain balance. A frequent error occurs when the inverse operation is performed on only one side, which results in an unbalanced equation. Another common mistake is the misapplication of the inverse process by neglecting to reverse the sign of the constant. For instance, in the equation
$$x - 3 = 7,$$
the correct procedure requires addition of 3 to both sides rather than subtraction. Such errors can be minimized by systematically identifying the operation that needs to be reversed and then carrying out the corresponding inverse operation on both sides of the equation.

Multiple Choice Questions

1. Which of the following best describes a one-step equation?

    (a) An equation requiring multiple sequential operations.

    (b) An equation that can be solved by applying a single inverse operation.

    (c) An equation that involves variables on both sides.

    (d) An equation with a variable only in the denominator.

2. In the equation
$$x + 8 = 15,$$
   which operation should be performed to isolate x?

    (a) Add 8 to both sides.

    (b) Multiply both sides by 8.

    (c) Subtract 8 from both sides.

    (d) Divide both sides by 8.

3. To solve the equation
$$5x = 25,$$
   which inverse operation is applied?

    (a) Multiply both sides by 5.

    (b) Divide both sides by 5.

    (c) Subtract 5 from both sides.

    (d) Add 5 to both sides.

4. In the equation
$$\frac{x}{7} = 3,$$
   what operation correctly isolates x?

    (a) Multiply both sides by 7.

    (b) Divide both sides by 7.

    (c) Add 7 to both sides.

    (d) Subtract 7 from both sides.

5. Which step is essential when using an inverse operation to solve a one-step equation?

    (a) Performing the operation on the variable side only.
    (b) Applying the inverse operation to both sides of the equation.
    (c) Ignoring the constant on one side.
    (d) Reapplying the original operation after isolating the variable.

6. Solve the equation
$$x - 4 = 10.$$

    (a) x = 14
    (b) x = 6
    (c) x = 10
    (d) x = 4

7. Which of the following demonstrates the correct use of an inverse operation in solving a one-step equation?

    (a) For
    $$x + 5 = 12,$$
    adding 5 to both sides.
    (b) For
    $$3x = 27,$$
    dividing both sides by 3.
    (c) For
    $$x - 3 = 7,$$
    subtracting 3 from both sides.
    (d) For
    $$\frac{x}{2} = 4,$$
    dividing both sides by 2.

Answers:

1. **B: An equation that can be solved by applying a single inverse operation** Explanation: A one-step equation requires only one arithmetic operation (such as addition, subtraction, multiplication, or division) to isolate the variable.

2. **C: Subtract 8 from both sides** Explanation: Since the equation is in the form x + 8 = 15, subtracting 8 (the inverse of addition) from both sides isolates x.

3. **B: Divide both sides by 5** Explanation: In the equation 5x = 25, x is multiplied by 5, so dividing both sides by 5 (the inverse of multiplication) isolates x.

4. **A: Multiply both sides by 7** Explanation: In $\frac{x}{7} = 3$, the variable x is divided by 7; multiplying both sides by 7 reverses the division and isolates x.

5. **B: Applying the inverse operation to both sides of the equation** Explanation: To maintain the balance of an equation, any operation (including an inverse operation) must be applied equally to both sides.

6. **A: x = 14** Explanation: For x - 4 = 10, adding 4 (the inverse of subtraction) to both sides yields x = 14.

7. **B: For $3x = 27$, dividing both sides by 3** Explanation: Dividing both sides by 3 correctly applies the inverse operation to cancel the multiplication by 3. The other options either apply the wrong operation or do not correctly isolate the variable.

Practice Problems

1. Solve the following one-step equation:

$$x + 5 = 12$$

2. Solve the following one-step equation:
$$x - 4 = 9$$

3. Solve the following one-step equation:
$$7x = 35$$

4. Solve the following one-step equation:
$$\frac{x}{6} = 4$$

5. Solve the following one-step equation:
$$2x = 16$$

6. Solve the following one-step equation:
$$x - 10 = 22$$

Answers

1. Solve the equation:
$$x + 5 = 12$$

    Solution:
    This equation involves an addition operation with the variable x. To isolate x, we subtract 5 from both sides (since subtraction is the inverse of addition):
    $$x + 5 - 5 = 12 - 5$$
    $$x = 7$$

Therefore, the solution is
$$x = 7.$$

2. Solve the equation:
$$x - 4 = 9$$

   Solution:
   Here, x is being subtracted by 4. To undo the subtraction, add 4 to both sides:
   $$x - 4 + 4 = 9 + 4$$
   $$x = 13$$

   Thus, the solution is
   $$x = 13.$$

3. Solve the equation:
$$7x = 35$$

   Solution:
   In this equation, x is multiplied by 7. The inverse operation is division. Dividing both sides by 7 gives:
   $$\frac{7x}{7} = \frac{35}{7}$$
   $$x = 5$$

   Hence, the solution is
   $$x = 5.$$

4. Solve the equation:
$$\frac{x}{6} = 4$$

   Solution:
   Here, x is divided by 6. To isolate x, multiply both sides by 6 (using the inverse operation of division):
   $$\frac{x}{6} \times 6 = 4 \times 6$$
   $$x = 24$$

   Therefore, the solution is
   $$x = 24.$$

5. Solve the equation:
$$2x = 16$$

Solution:
In this case, the variable x is multiplied by 2. To solve for x, divide both sides by 2:
$$\frac{2x}{2} = \frac{16}{2}$$
$$x = 8$$

So, the solution is
$$x = 8.$$

6. Solve the equation:
$$x - 10 = 22$$

Solution:
The equation shows that 10 is subtracted from x. To isolate x, add 10 to both sides:
$$x - 10 + 10 = 22 + 10$$
$$x = 32$$

Thus, the solution is
$$x = 32.$$

Chapter 28

Solving Multi-Step Equations

Recognizing Multi-Step Equations

Multi-step equations are algebraic statements that require a series of inverse operations in order to isolate the unknown variable. These equations often present variables accompanied by multiple arithmetic operations, such as distribution, addition, subtraction, multiplication, and division. In many cases, terms appear on both sides of the equality sign, and the equation must be simplified by regrouping and combining like terms. The proper sequence of operations, performed in reverse order of the standard order of operations, is essential for maintaining the balance of the equation while reducing its complexity.

Combining Like Terms and Eliminating Parentheses

When an equation includes parentheses, the first step involves the application of the distributive property in order to remove these grouping symbols. The distributive property states that

$$a(b + c) = ab + ac.$$

This property permits the expansion of expressions so that terms can be arranged appropriately. After distribution, like terms—that

is, terms that contain the same variable components—can be combined by adding or subtracting their coefficients. For example, an equation in the form
$$2(x+3) - x = 7$$
can be transformed by first distributing the multiplication:
$$2x + 6 - x = 7.$$
The similar terms $2x$ and $-x$ are then combined to yield:
$$x + 6 = 7.$$
This process of eliminating parentheses and merging like terms serves to streamline the equation, making the subsequent use of inverse operations more straightforward.

Applying Inverse Operations Sequentially

Once the equation has been simplified through the removal of parentheses and the combination of like terms, the next phase involves the application of inverse operations in a sequential manner. Each arithmetic operation applied to the variable must be reversed in order to isolate the variable step by step. For instance, when an equation is formatted as
$$ax + b = c,$$
the procedure begins by eliminating the constant term b. This is achieved by subtracting b from both sides:
$$ax + b - b = c - b,$$
which simplifies to
$$ax = c - b.$$
Following the removal of the constant term, the coefficient a attached to the variable x is eliminated by dividing both sides by a:
$$x = \frac{c-b}{a}.$$
In instances where variables appear on both sides of the equation, transferring all terms containing the variable to one side and the constant terms to the opposite side is a logical strategy. Each step, whether adding or subtracting a term or multiplying or dividing by a constant, must be executed on both sides of the equation to preserve the balance inherent in the equality.

Worked Examples

Example 1

Examine the equation
$$3(x+2) - 4 = 2x + 5.$$
The first operation applies the distributive property:
$$3x + 6 - 4 = 2x + 5.$$
Next, the constant terms on the left are combined:
$$3x + 2 = 2x + 5.$$
To consolidate the variable terms, subtract $2x$ from both sides:
$$3x - 2x + 2 = 2x - 2x + 5,$$
which simplifies to
$$x + 2 = 5.$$
Finally, subtract 2 from both sides to isolate x:
$$x = 5 - 2,$$
resulting in
$$x = 3.$$

Example 2

Consider the equation
$$2(2x - 3) + 4 = 3x + 1.$$
Begin by distributing the multiplication across the parentheses:
$$4x - 6 + 4 = 3x + 1.$$
Following distribution, combine the constant terms on the left:
$$4x - 2 = 3x + 1.$$
Subtract $3x$ from both sides to regroup the variable terms:
$$4x - 3x - 2 = 3x - 3x + 1,$$

resulting in
$$x - 2 = 1.$$

Finally, add 2 to both sides to isolate x:
$$x = 1 + 2,$$

which simplifies to
$$x = 3.$$

Example 3

Examine an equation where variable terms appear on both sides:
$$5x - 7 = 2x + 8.$$

Begin by isolating the variable terms on one side by subtracting $2x$ from both sides:
$$5x - 2x - 7 = 2x - 2x + 8,$$

which simplifies to
$$3x - 7 = 8.$$

Next, eliminate the constant term on the side containing the variable by adding 7 to both sides:
$$3x - 7 + 7 = 8 + 7,$$

resulting in
$$3x = 15.$$

Finally, divide both sides by 3 to isolate x:
$$x = \frac{15}{3},$$

yielding
$$x = 5.$$

Multiple Choice Questions

1. Which of the following best describes a multi-step equation?

    (a) An equation that can be solved in one simple operation.

    (b) An equation that involves only one arithmetic operation.

(c) An equation that requires several inverse operations to isolate the unknown variable.

(d) An equation that always contains only whole numbers.

2. When simplifying an equation containing parentheses, which property is used to eliminate the parentheses?

   (a) Associative Property
   (b) Commutative Property
   (c) Distributive Property
   (d) Identity Property

3. What is the correct first step for solving the equation
$$2(x+3) - x = 7?$$

   (a) Combine like terms before doing any other operation.
   (b) Distribute the 2 across the parentheses.
   (c) Subtract x from both sides immediately.
   (d) Add 3 to both sides.

4. In the equation
$$3(x+2) - 4 = 2x + 5,$$

   after expanding the parentheses and combining like terms, what should be the next step?

   (a) Add 4 to both sides.
   (b) Subtract $2x$ from both sides.
   (c) Multiply both sides by 3.
   (d) Divide both sides by 2.

5. For an equation of the form
$$ax + b = c,$$

   which operation is applied last to solve for x?

   (a) Dividing both sides by a.
   (b) Subtracting b from both sides.
   (c) Multiplying both sides by a.

(d) Adding b to both sides.

6. Why is it crucial to perform the same arithmetic operation on both sides of an equation?

    (a) It guarantees that the equation is written in a standard form.

    (b) It maintains the balance of the equation.

    (c) It eliminates fractional coefficients.

    (d) It automatically isolates the variable.

7. In the equation
$$2(2x - 3) + 4 = 3x + 1,$$
after applying the distributive property and combining like terms to obtain
$$4x - 2 = 3x + 1,$$
what is the correct next step to continue solving for x?

    (a) Subtract $3x$ from both sides to get $x - 2 = 1$.

    (b) Add 2 to both sides to get $4x = 3x + 3$.

    (c) Divide both sides by 2 immediately.

    (d) Multiply both sides by 3.

Answers:

1. **C: An equation that requires several inverse operations to isolate the unknown variable**
This is correct because multi-step equations involve steps like removing parentheses, combining like terms, and applying inverse operations (such as addition/subtraction and multiplication/division) to solve for the variable.

2. **C: Distributive Property**
The distributive property is used to eliminate parentheses by multiplying the term outside the parentheses with each term inside, allowing the equation to be simplified.

3. **B: Distribute the 2 across the parentheses**
The first step is to apply the distributive property in order to remove the parentheses, turning $2(x + 3)$ into $2x + 6$, which then allows you to combine like terms.

4. **B: Subtract $2x$ from both sides**
   After eliminating the parentheses and combining like terms, subtracting $2x$ from both sides groups the variable terms together, making it easier to isolate x.

5. **A: Dividing both sides by a**
   Once the equation is simplified to $ax = c - b$, dividing both sides by a is the final operation to isolate x.

6. **B: It maintains the balance of the equation**
   Applying the same operations to both sides of an equation preserves the equality. This property is fundamental to solving equations correctly.

7. **A: Subtract $3x$ from both sides to get $x - 2 = 1$**
   In this equation, after simplification, subtracting $3x$ from both sides isolates the x-term on one side, which is the correct step before adding 2 to both sides to solve for x.

Practice Problems

1. Solve the following multi-step equation:
$$3(x - 2) + 4 = 2x + 10$$

2. Solve the following multi-step equation:
$$5(x + 3) - 7 = 2(2x + 1)$$

3. Solve the following multi-step equation:
$$2(3x - 2) - (x + 4) = 3(x - 1) + 2$$

4. Solve the following multi-step equation:
$$4 - 2(3 - x) = 3x + 1$$

5. Solve the following multi-step equation:
$$3(2x + 4) - 5x = 2x + 7$$

6. Solve the following multi-step equation:
$$2(x+3) + 3(2-x) = 4(x-1) + 5$$

Answers

1. For the equation:
$$3(x-2) + 4 = 2x + 10$$

   Solution:
   First, distribute 3 on the left-hand side:
   $$3x - 6 + 4 = 2x + 10$$

   Combine like terms:
   $$3x - 2 = 2x + 10$$

   Subtract 2x from both sides:
   $$x - 2 = 10$$

   Finally, add 2 to both sides to isolate the variable:
   $$x = 12$$

   Therefore, the solution is **x = 12**.

2. For the equation:
$$5(x+3) - 7 = 2(2x+1)$$

Solution:
Distribute on both sides. On the left-hand side:
$$5x + 15 - 7 = 5x + 8$$
On the right-hand side:
$$4x + 2$$
Set the expressions equal:
$$5x + 8 = 4x + 2$$
Subtract 4x from both sides:
$$x + 8 = 2$$
Then, subtract 8 from both sides:
$$x = -6$$
Thus, the solution is **x = -6**.

3. For the equation:
$$2(3x - 2) - (x + 4) = 3(x - 1) + 2$$
Solution:
Start by expanding each term:
$$6x - 4 - x - 4 = 3x - 3 + 2$$
Simplify the left-hand side by combining like terms:
$$5x - 8$$
And simplify the right-hand side:
$$3x - 1$$
Now, the equation is:
$$5x - 8 = 3x - 1$$
Subtract 3x from both sides:
$$2x - 8 = -1$$

Add 8 to both sides:
$$2x = 7$$
Divide both sides by 2:
$$x = \frac{7}{2}$$
Therefore, the solution is **x = 7/2**.

4. For the equation:
$$4 - 2(3 - x) = 3x + 1$$
   Solution:
   Begin by distributing -2 on the left-hand side:
$$4 - 6 + 2x = 3x + 1$$
   Simplify the left-hand side:
$$2x - 2 = 3x + 1$$
   Subtract 2x from both sides to collect like terms:
$$-2 = x + 1$$
   Then, subtract 1 from both sides:
$$x = -3$$
   Thus, the solution is **x = -3**.

5. For the equation:
$$3(2x + 4) - 5x = 2x + 7$$
   Solution:
   First, distribute 3 on the left-hand side:
$$6x + 12 - 5x = x + 12$$
   Now, set the equation:
$$x - 12 = 2x + 7$$
   Subtract x from both sides:
$$12 = x + 7$$
   Finally, subtract 7 from both sides:
$$x = 5$$
   Therefore, the solution is **x = 5**.

6. For the equation:
$$2(x+3) + 3(2-x) = 4(x-1) + 5$$

Solution:
Begin by distributing on the left-hand side:
$$2x + 6 + 6 - 3x = -x + 12$$

Next, distribute on the right-hand side:
$$4x - 4 + 5 = 4x + 1$$

Now, write the simplified equation:
$$-x + 12 = 4x + 1$$

Add x to both sides:
$$12 = 5x + 1$$

Subtract 1 from both sides:
$$11 = 5x$$

Divide both sides by 5:
$$x = \frac{11}{5}$$

Thus, the solution is **x = 11/5**.

Chapter 29

Equations with Variables on Both Sides

Understanding the Structure of the Equation

Linear equations in which the unknown appears on both sides involve terms that comprise variables and constants distributed throughout the equality. In these equations, each side may contain several terms that must be manipulated so that all variable terms are consolidated on one side, while the constant terms are grouped on the opposite side. The principle of balance governs the process: any operation performed on one side of the equals sign must be equally applied to the other side. This section describes the framework within which the equation is constructed and prepares the groundwork for the systematic approach to solving the equation.

Transposing Variable Terms

To simplify an equation when the variable appears on both sides, the initial step is to move all terms containing the variable to one side of the equation. This is achieved by using additive inverse operations. For example, if an equation contains the term 2x on one side, subtracting 2x from both sides will eliminate it from one side, thereby maintaining the balance of the equation. The decision regarding which side to collect the variable terms on is arbitrary; however, selecting the side with the larger coefficient or the fewest negative signs may simplify subsequent computations. Through the process of transposition, the structure of the equation is rearranged into a form that is more conducive to the combination of like terms.

Combining Like Terms and Isolating the Variable

After transposing the variable terms, the next step involves combining like terms to reach a simplified form that clearly delineates the variable component. Constant terms are gathered and simplified on the opposite side by using addition or subtraction. Once the equation is rewritten with the variable part and the constant part separated, inverse operations are applied to isolate the variable. For instance, an equation that reduces to the format $ax + c = d$ can be simplified by subtracting the constant c from both sides to attain $ax = d - c$. Finally, dividing both sides by the coefficient a isolates the variable completely. This systematic process reinforces the fundamental property that equitable operations on both sides preserve the balance of the equality.

Strategies for Avoiding Common Errors

Maintaining the integrity of an equation during manipulation is critical. Every term that is moved from one side to the other must have its sign appropriately altered, and any operation—whether it involves addition, subtraction, multiplication, or division—must be performed equally on both sides of the equation. A frequent error occurs when a term is transposed without changing its sign, which disrupts the equality. Another common pitfall is the failure to combine like terms correctly before proceeding with the isolation of the variable. These errors can be avoided by carefully rewriting each step before executing the next operation. Emphasis on careful and deliberate manipulation ensures the balance remains intact and that the process of solving the equation is both methodical and error-free.

Worked Examples

Consider the equation below, which contains the variable on both sides:
$$3x + 4 = 2x + 6.$$

First, subtract 2x from both sides to collect the variable terms on one side:
$$3x - 2x + 4 = 2x - 2x + 6,$$
which simplifies to
$$x + 4 = 6.$$

Next, subtract 4 from both sides to isolate the variable:
$$x + 4 - 4 = 6 - 4,$$
yielding the solution
$$x = 2.$$

A different example illustrates the process further:
$$5x - 7 = 2x + 8.$$

Begin by subtracting 2x from both sides to move all x-terms to the left:
$$5x - 2x - 7 = 2x - 2x + 8,$$
resulting in
$$3x - 7 = 8.$$

Then, add 7 to both sides to eliminate the constant on the left:
$$3x - 7 + 7 = 8 + 7,$$
which simplifies to
$$3x = 15.$$

Finally, divide both sides by 3 to isolate x:

$$x = \frac{15}{3},$$

yielding the solution

$$x = 5.$$

Each of these examples demonstrates the systematic approach of transposing variable terms, combining like terms, and applying inverse operations. The detailed manipulation of each term underpins the fundamental concept of maintaining balance throughout the process.

Multiple Choice Questions

1. Which of the following best describes an equation with variables on both sides?

    (a) An equation that contains only variable terms on one side.

    (b) An equation with no variable terms.

    (c) An equation where variable terms appear on both sides along with constant terms.

    (d) An equation that contains only constant terms.

2. What does the principle of balance in an equation mean?

    (a) Performing an operation on only one side of the equation.

    (b) Ensuring that both sides have an equal number of terms.

    (c) Applying the same operation to both sides of the equation to maintain equality.

    (d) Changing the sign of a term on one side without altering the other side.

3. When transposing a variable term from one side of an equation to the other, what should you do with that term?

    (a) Move it as is, keeping the same sign.

    (b) Change its sign by adding its additive inverse before moving it.

(c) Multiply it by -1 and then add it without further modification.

(d) Divide it by the variable's coefficient before moving it.

4. Combining like terms in an equation is important because it:

    (a) Increases the number of terms, making the equation more detailed.

    (b) Simplifies the equation by reducing similar terms, thereby easing the isolation of the variable.

    (c) Introduces additional variables into the equation.

    (d) Changes the equality by shifting terms randomly.

5. After gathering all variable terms on one side and constant terms on the other, what is the next step to solve for the variable?

    (a) Multiply both sides by the variable.

    (b) Subtract the variable term from both sides.

    (c) Divide both sides by the coefficient of the variable.

    (d) Add the coefficients together.

6. Which of the following is a common error when transposing terms in an equation with variables on both sides?

    (a) Adding the same term to both sides correctly.

    (b) Forgetting to change the sign of the term that is being transposed.

    (c) Dividing both sides by a nonzero number.

    (d) Multiplying both sides by the coefficient of the variable.

7. Consider the equation: $5x - 7 = 2x + 8$. Which of the following is the correct solution for x?

    (a) x = 1

    (b) x = 3

    (c) x = 5

    (d) x = -5

Answers:

1. **C: An equation where variable terms appear on both sides along with constant terms.** This option correctly describes an equation in which variables are located on both sides, requiring the transposition of terms and subsequent simplification to solve for the variable.

2. **C: Applying the same operation to both sides of the equation to maintain equality.** The principle of balance means that any operation performed on one side of an equation must also be performed on the other side to preserve the equality.

3. **B: Change its sign by adding its additive inverse before moving it.** When transposing a term, you must change its sign (for instance, moving +2x becomes -2x on the other side) to maintain the balance of the equation.

4. **B: Simplifies the equation by reducing similar terms, thereby easing the isolation of the variable.** Combining like terms minimizes the number of terms in the equation, making it clearer and simpler to isolate the variable.

5. **C: Divide both sides by the coefficient of the variable.** Once the equation is simplified to a form like ax = b, dividing both sides by the coefficient a isolates the variable and provides the solution.

6. **B: Forgetting to change the sign of the term that is being transposed.** A common mistake is moving a term from one side to the other without changing its sign, which disrupts the balance of the equation and leads to incorrect results.

7. **C: x = 5** To solve 5x - 7 = 2x + 8, first subtract 2x from both sides to get 3x - 7 = 8, then add 7 to both sides to obtain 3x = 15, and finally divide both sides by 3 to find x = 5.

Practice Problems

1. Solve the equation:

$$3x + 4 = 2x + 6.$$

2. Solve the equation:
$$5x - 7 = 2x + 8.$$

3. Solve the equation:
$$4x + 9 = 7x - 6.$$

4. Solve the equation:
$$2(x + 5) = 3x - 4.$$

5. Solve the equation:
$$6 - 2x = 4x + 10.$$

6. Solve the equation:
$$3(2x - 1) = 2(x + 4) + x.$$

Answers

1. For the equation
$$3x + 4 = 2x + 6,$$
begin by subtracting 2x from both sides to consolidate the variable terms:
$$3x - 2x + 4 = 6.$$
This simplifies to:
$$x + 4 = 6.$$
Next, subtract 4 from both sides to isolate the variable:
$$x = 6 - 4 = 2.$$
Therefore, the solution is
$$x = 2.$$

2. For the equation
$$5x - 7 = 2x + 8,$$
subtract 2x from both sides to group the variable terms on one side:
$$5x - 2x - 7 = 8.$$
This gives:
$$3x - 7 = 8.$$
Then, add 7 to both sides:
$$3x = 8 + 7 = 15.$$
Finally, divide both sides by 3:
$$x = \frac{15}{3} = 5.$$
Thus, the solution is
$$x = 5.$$

3. For the equation
$$4x + 9 = 7x - 6,$$
subtract 4x from both sides to start gathering like terms:
$$9 = 7x - 4x - 6.$$
Simplify the right-hand side:
$$9 = 3x - 6.$$
Next, add 6 to both sides:
$$15 = 3x.$$
Finally, divide both sides by 3:
$$x = \frac{15}{3} = 5.$$
Therefore, the solution is
$$x = 5.$$

4. For the equation
$$2(x + 5) = 3x - 4,$$
begin by applying the distributive property to the left-hand side:
$$2x + 10 = 3x - 4.$$
Next, subtract 2x from both sides:
$$10 = 3x - 2x - 4,$$
which simplifies to:
$$10 = x - 4.$$
Then, add 4 to both sides to isolate x:
$$x = 10 + 4 = 14.$$
Thus, the solution is
$$x = 14.$$

5. For the equation
$$6 - 2x = 4x + 10,$$
subtract 4x from both sides to collect the x-terms on one side:
$$6 - 2x - 4x = 10.$$
This simplifies to:
$$6 - 6x = 10.$$
Next, subtract 6 from both sides:
$$-6x = 10 - 6,$$
or
$$-6x = 4.$$
Finally, divide by -6:
$$x = \frac{4}{-6} = -\frac{2}{3}.$$
Therefore, the solution is
$$x = -\frac{2}{3}.$$

6. For the equation
$$3(2x - 1) = 2(x + 4) + x,$$
first distribute on both sides. For the left-hand side:
$$3 \cdot 2x - 3 \cdot 1 = 6x - 3.$$
For the right-hand side, distribute and combine like terms:
$$2x + 8 + x = 3x + 8.$$
The equation now reads:
$$6x - 3 = 3x + 8.$$
Subtract 3x from both sides to group the variable terms:
$$3x - 3 = 8.$$

Then, add 3 to both sides:
$$3x = 8 + 3 = 11.$$

Finally, divide by 3:
$$x = \frac{11}{3}.$$

Thus, the solution is
$$x = \frac{11}{3}.$$

Chapter 30

Introduction to Inequalities

Definition and Fundamental Concepts

An inequality is a mathematical statement that expresses the relationship between two expressions which are not necessarily equal. Instead of an equals sign, an inequality employs comparison symbols to denote that one expression is either less than, greater than, less than or equal to, or greater than or equal to another expression. In algebraic contexts, these symbols—namely, $<$, $>$, $\leq$, and $\geq$—establish relationships that can be used to describe ranges of values rather than pinpoint a unique solution. The formulation of an inequality involves the intrinsic idea that the quantities in question assume an ordered structure, which is pivotal in understanding the behavior of functions and expressions algebraically.

Inequality Symbols and Their Meanings

The symbol "$<$" indicates that the expression on the left is strictly less than the one on the right, while "$>$" signifies that the left-hand expression is strictly greater than the right-hand one. The symbols "$\leq$" and "$\geq$" expand these comparisons by incorporating the possibility of equality. In the expression $a \leq b$, the value of a may be either less than or equal to b, and similarly, $a \geq b$ indicates that a is either greater than or equal to b. These symbols serve as the building blocks for constructing and interpreting inequalities, offering a concise way to communicate numerical relationships and constraints algebraically.

Algebraic Techniques for Solving Inequalities

The process for solving inequalities is closely aligned with the methods used for solving equations, yet it includes additional considerations dictated by the nature of the inequality. Initially, similar terms are combined to simplify the expression. Maintaining the balance of an inequality requires that any algebraic operation performed on one side be equally applied to the other side. When the same nonnegative value is added to or subtracted from both sides, the direction of the inequality remains unchanged. However, multiplying or dividing both sides by a negative number necessitates reversing the inequality sign in order to preserve the truth of the statement. For example, if an inequality is given by $-ax > b$ with a positive, dividing both sides by $-a$ leads to the reversed inequality $x < -\frac{b}{a}$. This property is essential in the isolation of the variable, ensuring that the solution set accurately reflects the range of values that satisfy the original inequality.

Representation of Solution Sets

Once an inequality has been solved algebraically, the solution can be expressed in multiple forms. One common method is interval notation, which succinctly describes the exact range of values that satisfy the inequality. A solution such as $x \leq 5$ may be represented as $(-\infty, 5]$, where the bracket indicates that 5 is included in the solution set. Alternatively, the solution may be depicted on a number line, where a ray or a shaded region visually demonstrates all the real numbers that fulfill the inequality. This graphical representation provides an intuitive understanding of how the solution spans a continuum of values rather than a single fixed number.

Worked Examples

Consider the inequality
$$3x - 5 \leq 10.$$

The first step involves isolating the term containing the variable by adding 5 to both sides, yielding

$$3x \leq 15.$$

Subsequently, dividing both sides by 3 leads to

$$x \leq 5,$$

which indicates that the set of solutions includes all real numbers that are less than or equal to 5.

Another example employs an inequality where the variable is associated with a negative coefficient. Take the inequality

$$-2x > 8.$$

Dividing both sides by -2 while reversing the direction of the inequality produces

$$x < -4,$$

thus identifying all real numbers strictly less than -4 as the solution.

Multiple Choice Questions

1. Which of the following best defines an inequality?

    (a) A statement that shows two expressions are identical.

    (b) A statement that compares two expressions using symbols such as $<$, $>$, $\leq$, or $\geq$.

    (c) A statement that applies only to whole numbers.

    (d) A statement that can only be solved by performing addition.

2. Which inequality symbol is used to denote "less than or equal to"?

    (a) $<$

    (b) $>$

    (c) $\leq$

    (d) $\geq$

3. When solving an inequality, what must you do if you multiply or divide both sides by a negative number?

   (a) Leave the inequality sign unchanged.
   (b) Reverse the inequality sign.
   (c) Multiply only the left side by the negative number.
   (d) Subtract a negative number from both sides.

4. Solve the inequality: $-4x < 8$. What is the correct solution for x?

   (a) $x < -2$
   (b) $x > -2$
   (c) $x < 2$
   (d) $x > 2$

5. Express the inequality $x \geq 3$ in interval notation.

   (a) $(3, \infty)$
   (b) $[3, \infty)$
   (c) $(-\infty, 3]$
   (d) $(-\infty, 3)$

6. How is the solution set for the inequality $x < 5$ correctly represented on a number line?

   (a) A closed circle at 5 with shading toward the left.
   (b) An open circle at 5 with shading toward the left.
   (c) A closed circle at 5 with shading toward the right.
   (d) An open circle at 5 with shading toward the right.

7. Which of the following statements about the properties of inequalities is true?

   (a) Adding or subtracting the same number from both sides does not change the inequality's direction.
   (b) Multiplying or dividing both sides by a positive number reverses the inequality sign.
   (c) Multiplying or dividing both sides by a negative number does not affect the inequality.

(d) Dividing by zero is a valid method for simplifying inequalities.

Answers:

1. **B: A statement that compares two expressions using symbols such as $<$, $>$, $\leq$, or $\geq$.**
   This is the correct definition of an inequality. It describes a relationship between two expressions where they are not necessarily equal, using comparison symbols.

2. **C: $\leq$**
   The symbol "$\leq$" denotes "less than or equal to," indicating that the first expression is either less than or exactly equal to the second.

3. **B: Reverse the inequality sign.**
   When you multiply or divide both sides of an inequality by a negative number, you must reverse the direction of the inequality sign to maintain the truth of the statement.

4. **B: $x > -2$**
   To solve $-4x < 8$, divide both sides by -4 (remember to reverse the inequality sign). This gives $x > -2$.

5. **B: $[3, \infty)$**
   The inequality $x \geq 3$ means that 3 is included in the set of solutions. In interval notation, a bracket is used at 3, leading to $[3, \infty)$.

6. **B: An open circle at 5 with shading toward the left.**
   Because $x < 5$ does not include 5, an open circle is used at 5, with shading toward the left to indicate all numbers less than 5.

7. **A: Adding or subtracting the same number from both sides does not change the inequality's direction.**
   This statement is true. Unlike multiplication or division by a negative number (which requires reversing the sign), adding or subtracting the same number from both sides of an inequality does not affect the direction of the inequality.

Practice Problems

1. Solve the inequality:
$$2x - 3 < 7$$

2. Solve the inequality:
$$-3(x - 2) \geq 6$$

3. Solve the compound inequality:
$$-3 < 2x + 1 \leq 7$$

4. Solve the inequality:
$$3 - 2x \geq x + 1$$

5. Graph the inequality and express the solution in interval notation:
$$x > 4$$

6. Write and solve the inequality from the following statement: "A number decreased by 5 is greater than or equal to 10."
$$x - 5 \geq 10$$

Answers

1. **Solution:**
   We start with the inequality:
   $$2x - 3 < 7$$
   Add 3 to both sides to isolate the term with x:
   $$2x < 10$$
   Now, divide both sides by 2:
   $$x < 5$$
   Explanation: All real numbers less than 5 satisfy the inequality. In interval notation, the solution set is $(-\infty, 5)$.

2. **Solution:**
   Begin with:
   $$-3(x - 2) \geq 6$$
   First, distribute -3 across the parenthesis:
   $$-3x + 6 \geq 6$$
   Subtract 6 from both sides:
   $$-3x \geq 0$$
   Divide both sides by -3. Remember, dividing by a negative number reverses the inequality sign:
   $$x \leq 0$$
   Explanation: The solution set consists of all real numbers less than or equal to 0, represented in interval notation as $(-\infty, 0]$.

3. **Solution:**
   Start with the compound inequality:
   $$-3 < 2x + 1 \leq 7$$
   Subtract 1 from all three parts to isolate the term with x:
   $$-4 < 2x \leq 6$$

Now, divide each part by 2:
$$-2 < x \leq 3$$

Explanation: The solution requires x to be greater than -2 and less than or equal to 3. In interval notation, this is expressed as $(-2, 3]$.

4. **Solution:**
   Consider the inequality:
   $$3 - 2x \geq x + 1$$
   Subtract x from both sides to combine like terms:
   $$3 - 3x \geq 1$$
   Next, subtract 3 from both sides:
   $$-3x \geq -2$$
   Divide both sides by -3, and remember to reverse the inequality sign:
   $$x \leq \frac{2}{3}$$
   Explanation: The final solution set includes all values of x that are less than or equal to $\frac{2}{3}$.

5. **Solution:**
   The inequality is:
   $$x > 4$$
   Explanation: This inequality represents all real numbers strictly greater than 4. Graphically, you would draw an open circle at 4 on the number line and shade the region to the right. The solution in interval notation is:
   $$(4, \infty)$$

6. **Solution:**
   First, translate the statement "A number decreased by 5 is greater than or equal to 10" into an inequality:
   $$x - 5 \geq 10$$

Next, add 5 to both sides:

$$x \geq 15$$

Explanation: The solution indicates that any number x that is 15 or greater satisfies the condition. In interval notation, the solution set is given by:

$$[15, \infty)$$

Chapter 31

Solving Multi-Step Inequalities

Fundamental Concepts and Properties

Mathematical inequalities serve to express relationships that are not strict equalities between two algebraic expressions. In multi-step inequalities, several operations must be carried out sequentially in order to isolate the variable. The same basic arithmetic properties that are used when solving equations apply here; however, particular care is required when an operation involves a negative factor. In such cases, the direction of the inequality must be reversed in order to maintain the validity of the statement. The properties of addition, subtraction, multiplication, and division extend to inequalities provided these rules are followed closely.

In any inequality, if the same number is added to or subtracted from each side, the relative order of the expressions remains unchanged. On the other hand, multiplying or dividing both sides of an inequality by a positive number preserves the inequality's direction, while performing these operations with a negative number necessitates a reversal of the inequality sign. This principle is intrinsic to solving multi-step inequalities and ensures that the resulting solution set accurately reflects the range of values for which the original inequality holds true.

Procedure for Isolating the Variable

The process of isolating the variable in multi-step inequalities begins with simplifying both sides of the inequality by combining like terms and distributing any factors over grouped terms. The objective is to create an expression in which the variable term is isolated on one side of the inequality sign. Once like terms have been combined, operations such as addition or subtraction are applied across the inequality to further group the variable terms on one side and the constant terms on the other.

One common strategy involves repositioning the variable terms to one side by subtracting or adding them to both sides of the inequality, and similarly repositioning the constant terms. After this rearrangement, an operation that involves multiplication or division is typically required to solve for the variable. Special attention must be given if the multiplication or division involves a negative number, as the inequality sign must be reversed to maintain an accurate representation of the relationship.

Combining Like Terms and Distributing Factors

A critical step in solving multi-step inequalities is the simplification of expressions through the combination of like terms and the careful distribution of factors. This involves using the distributive property to eliminate parentheses and then combining coefficients of similar variables. Such rearrangements are essential for reducing complex expressions into a more tractable form.

For instance, when an inequality includes terms such as a variable multiplied by a constant that is outside parentheses, distributing that constant across each term inside the parentheses produces an expression that can be simplified by combining like terms. Performing these operations systematically minimizes the chances of errors and ensures that every step adheres to the established properties of inequalities.

Techniques Involving Negative Multiplication or Division

In the course of simplifying an inequality, a situation may arise where the variable must be isolated through multiplication or division by a negative number. It is imperative that during such operations the direction of the inequality is reversed. This rule exists to preserve the ordered structure that the inequality represents. For example, in an inequality where the expression $-ax \geq b$ is encountered with a being a positive constant, division by $-a$ must be accompanied by the reversal of the inequality sign, resulting in an inequality of the form $x \leq -\frac{b}{a}$.

The reversal condition is a fundamental property that differentiates the process of solving inequalities from that of solving equalities. Each instance of multiplication or division by a negative factor requires a deliberate and careful adjustment of the inequality symbol so that the resulting solution remains valid.

Worked Example of a Multi-Step Inequality

Consider the inequality
$$-2(3x - 4) + 6 \leq 10 + x.$$

The initial step involves distributing the factor -2 across the terms within the parentheses, yielding
$$-6x + 8 + 6 \leq 10 + x.$$

The next step combines the constant terms on the left side to produce
$$-6x + 14 \leq 10 + x.$$

Subsequently, subtracting x from both sides groups the variable terms together:
$$-7x + 14 \leq 10.$$

Continuing with the isolation of the variable, subtract 14 from both sides to obtain
$$-7x \leq -4.$$

At this stage, dividing both sides by -7 is necessary. Since division is by a negative number, the inequality sign must be reversed, resulting in
$$x \geq \frac{4}{7}.$$
This example embodies the step-by-step approach that is characteristic of solving multi-step inequalities, with careful attention paid to each arithmetic operation and the properties that govern them.

Graphical Representation and Interval Notation

Upon obtaining a solution for a multi-step inequality, the resulting set of values can be represented using interval notation and/or illustrated on a number line. In interval notation, the endpoint is enclosed within brackets if the inequality is nonstrict (featuring symbols such as $\geq$ or $\leq$) and enclosed within parentheses if the endpoint is excluded by a strict inequality (such as $<$ or $>$). In addition, a graphical representation on a number line provides a visual depiction of the solution set. A corresponding point is marked on the number line with either an open or closed circle depending on whether the endpoint is included in the solution, and a ray or shaded region extends from the marked point to indicate the range of values that satisfy the inequality.

The clarity achieved by combining symbolic and graphical representations reinforces the techniques employed in solving multi-step inequalities, ensuring that every transformation remains consistent with the algebraic properties of inequalities.

Multiple Choice Questions

1. When multiplying or dividing both sides of an inequality by a negative number, you must:

    (a) Leave the inequality sign unchanged.

    (b) Reverse the direction of the inequality sign.

    (c) Multiply the other side by -1.

    (d) Change the inequality into an equation.

2. What is the correct first step in solving a multi-step inequality?

    (a) Isolate the variable immediately.
    (b) Distribute any factors and combine like terms.
    (c) Multiply both sides by the variable.
    (d) Add the same constant to both sides without simplifying.

3. When simplifying an inequality, which operation requires you to flip the inequality sign?

    (a) Adding a number to both sides.
    (b) Subtracting a number from both sides.
    (c) Dividing both sides by a negative number.
    (d) Multiplying both sides by a positive number.

4. In the inequality
$$-2(3x - 4) + 6 \leq 10 + x,$$
which step should be performed immediately after distributing -2?

    (a) Subtract x from both sides.
    (b) Combine like constant terms.
    (c) Divide both sides by -2.
    (d) Reverse the inequality sign.

5. What is the purpose of moving all variable terms to one side and all constant terms to the other side when solving an inequality?

    (a) To eliminate the need for flipping the inequality sign.
    (b) To create an expression that is easier to solve.
    (c) To confuse the inequality.
    (d) To change the inequality into an equation.

6. Consider the inequality after combining like terms:
$$-7x \leq -4.$$
What is the correct solution for x after isolating the variable?

(a) $x \leq \frac{4}{7}$

(b) $x \geq \frac{4}{7}$

(c) $x \leq -\frac{4}{7}$

(d) $x \geq -\frac{4}{7}$

7. How is the solution set of a non-strict inequality (using $\leq$ or $\geq$) most appropriately represented on a number line?

    (a) With an open circle at the endpoint.

    (b) With a closed circle at the endpoint.

    (c) With a dashed line at the endpoint.

    (d) With a star marking the endpoint.

Answers:

1. **B: Reverse the direction of the inequality sign** When multiplying or dividing both sides of an inequality by a negative number, the inequality sign must be reversed to maintain the correct relationship between the expressions.

2. **B: Distribute any factors and combine like terms** The initial step in solving a multi-step inequality is to simplify each side of the inequality. This involves distributing any factors to eliminate parentheses and then combining like terms.

3. **C: Dividing both sides by a negative number** If you divide (or multiply) both sides of an inequality by a negative number, you must reverse the inequality sign to preserve the true relationship between the two sides.

4. **B: Combine like constant terms** After distributing -2 in the expression $-2(3x-4)+6$, the next step is to combine the constant terms. This simplification is essential before moving on to isolating the variable.

5. **B: To create an expression that is easier to solve** Rearranging the inequality so that variable terms are on one side and constant terms on the other simplifies the process of isolating the variable, which is crucial for finding the correct solution.

6. **B: $x \geq \frac{4}{7}$** Dividing both sides of $-7x \leq -4$ by -7 (and remembering to reverse the inequality sign because the division involves a negative number) results in $x \geq \frac{4}{7}$.

7. **B: With a closed circle at the endpoint** In a non-strict inequality (using $\leq$ or $\geq$), the endpoint is included in the solution set and is represented on a number line by a closed circle.

Practice Problems

1. Solve the inequality:
$$2(3x - 4) - 5 < 7x + 2.$$

2. Solve the inequality:
$$-3(2x + 1) + 4 \geq x - 5.$$

3. Solve the inequality:
$$4 - 2(3 - x) < 5x + 1.$$

4. Solve the inequality:
$$-5x + 3 < 2(1 - 3x) + 7.$$

5. Solve the inequality:
$$-4(2x - 3) + 5 \geq 3x - 6.$$

6. Solve the inequality:
$$2 - 3(1 - 2x) > -4(x - 2).$$

Answers

1. For the inequality
$$2(3x - 4) - 5 < 7x + 2,$$
begin by applying the distributive property on the left side:
$$6x - 8 - 5 < 7x + 2.$$
Combine like terms:
$$6x - 13 < 7x + 2.$$
Subtract 6x from both sides to isolate the variable term on the right:
$$-13 < x + 2.$$
Then subtract 2 from both sides:
$$-15 < x.$$
This inequality can be written in the form:
$$x > -15.$$
Thus, the solution is all x such that x > -15.

2. For the inequality
$$-3(2x + 1) + 4 \geq x - 5,$$
first distribute -3:
$$-6x - 3 + 4 \geq x - 5.$$
Combine the constant terms:
$$-6x + 1 \geq x - 5.$$
Subtract x from both sides to gather the x terms:
$$-7x + 1 \geq -5.$$
Subtract 1 from both sides:
$$-7x \geq -6.$$

Now, divide both sides by -7. Remember that when dividing by a negative number, the inequality sign must be reversed:

$$x \leq \frac{-6}{-7} \implies x \leq \frac{6}{7}.$$

Therefore, the solution is all x such that x is less than or equal to 6/7.

3. For the inequality

$$4 - 2(3 - x) < 5x + 1,$$

begin by distributing -2 on the left side:

$$4 - 6 + 2x < 5x + 1.$$

Combine like terms:

$$2x - 2 < 5x + 1.$$

Subtract 2x from both sides:

$$-2 < 3x + 1.$$

Then subtract 1 from both sides:

$$-3 < 3x.$$

Finally, divide both sides by 3:

$$-1 < x,$$

which is equivalent to:

$$x > -1.$$

So, the solution is all x such that x > -1.

4. For the inequality

$$-5x + 3 < 2(1 - 3x) + 7,$$

start by distributing 2 on the right side:

$$-5x + 3 < 2 - 6x + 7.$$

Combine the constant terms on the right:
$$-5x + 3 < -6x + 9.$$
Add 6x to both sides to bring the variable terms together:
$$x + 3 < 9.$$
Subtract 3 from both sides:
$$x < 6.$$
Therefore, the solution is all x such that x < 6.

5. For the inequality
$$-4(2x - 3) + 5 \geq 3x - 6,$$
first distribute -4 on the left side:
$$-8x + 12 + 5 \geq 3x - 6.$$
Combine the constants:
$$-8x + 17 \geq 3x - 6.$$
Add 8x to both sides:
$$17 \geq 11x - 6.$$
Then add 6 to both sides:
$$23 \geq 11x.$$
Finally, divide both sides by 11:
$$x \leq \frac{23}{11}.$$
Hence, the solution is all x such that x is less than or equal to 23/11.

6. For the inequality
$$2 - 3(1 - 2x) > -4(x - 2),$$
begin by distributing on both sides. First, distribute -3 on the left side:
$$2 - 3 + 6x = 6x - 1.$$

Next, distribute -4 on the right side:
$$-4x + 8.$$

Now, the inequality becomes:
$$6x - 1 > -4x + 8.$$

Add 4x to both sides to collect like terms:
$$10x - 1 > 8.$$

Add 1 to both sides:
$$10x > 9.$$

Divide both sides by 10:
$$x > \frac{9}{10}.$$

Thus, the solution is all x such that $x > \frac{9}{10}$.

Chapter 32

Exploring Absolute Value

Definition and Meaning of Absolute Value

The absolute value of a real number represents its distance from zero on the number line, regardless of the direction. In formal mathematical notation, the absolute value of a number x is denoted by $|x|$ and is defined by the piecewise function

$$|x| = \begin{cases} x, & \text{if } x \geq 0, \\ -x, & \text{if } x < 0. \end{cases}$$

This definition ensures that the result is always nonnegative. The concept emphasizes the numerical magnitude without concern for the sign. In this way, absolute value captures a fundamental measure of distance in the real number system, making it a useful tool to express how far a number lies from the origin.

Key Properties of Absolute Value

The absolute value function possesses several important properties that contribute to its utility in various mathematical contexts. A thorough understanding of these properties facilitates its application in both equations and inequalities.

1. **Non-Negativity:** For any real number x, the value $|x|$ is always greater than or equal to 0.

2. **Even Function:** The absolute value function is even, which means that $|x| = |-x|$ for all real numbers x. This reflects a symmetry about the vertical axis on the number line.

3. **Multiplicative Property:** The function satisfies the relation $|ab| = |a| \cdot |b|$ for any two real numbers a and b. This property is helpful in simplifying expressions and solving equations.

4. **Triangle Inequality:** An important and frequently applied property in various branches of mathematics is the triangle inequality, which states that for any real numbers a and b,
$$|a + b| \leq |a| + |b|.$$
This inequality describes how the absolute value of a sum is bounded by the sum of the absolute values.

Application in Equations

Absolute value equations are those in which the absolute value expression is set equal to a nonnegative constant. For an equation of the form
$$|A(x)| = B,$$
where $B \geq 0$, the definition of absolute value leads to two distinct cases:
$$A(x) = B \quad \text{or} \quad A(x) = -B.$$
For instance, in the equation
$$|2x - 3| = 5,$$
the two cases to consider are
$$2x - 3 = 5 \quad \text{and} \quad 2x - 3 = -5.$$
Solving the first equation by adding 3 to both sides yields
$$2x = 8 \quad \implies \quad x = 4.$$
Solving the second equation, adding 3 gives
$$2x = -2 \quad \implies \quad x = -1.$$
Thus, the equation $|2x - 3| = 5$ has two solutions, $x = 4$ and $x = -1$. In cases where the constant on the right-hand side is negative, the equation has no solutions because the absolute value cannot produce a negative result.

Application in Inequalities

Absolute value inequalities are useful for describing ranges of values that lie within a certain distance from a given point. Two common forms are those with the inequality symbol or $\leq$ and those with or $\geq$.

For an inequality of the form

$$|A(x)| < B,$$

with $B > 0$, the definition implies that the expression $A(x)$ must lie between $-B$ and B. This can be written as a compound inequality:

$$-B < A(x) < B.$$

As an example, consider the inequality

$$|x - 2| < 3.$$

This inequality implies that

$$-3 < x - 2 < 3.$$

Adding 2 to each part of the compound inequality results in

$$-1 < x < 5.$$

Thus, the solutions of the inequality are all real numbers x that satisfy the condition $-1 < x < 5$.

When faced with an inequality of the form

$$|A(x)| > B,$$

the situation differs because the requirement is that the expression $A(x)$ must lie outside the interval $[-B, B]$. This leads to two separate inequalities:

$$A(x) < -B \quad \text{or} \quad A(x) > B.$$

For example, the inequality

$$|3x + 1| \geq 7$$

can be split into the cases

$$3x + 1 \geq 7 \quad \text{or} \quad 3x + 1 \leq -7.$$

Solving the first inequality by subtracting 1 and then dividing by 3 results in
$$3x \geq 6 \implies x \geq 2.$$
The second inequality, upon subtracting 1 and dividing by 3, gives
$$3x \leq -8 \implies x \leq -\frac{8}{3}.$$
Hence, the solution set for the inequality $|3x + 1| \geq 7$ consists of all x values that satisfy either $x \geq 2$ or $x \leq -\frac{8}{3}$.

The process of rewriting absolute value inequalities into their equivalent compound inequalities and solving them systematically demonstrates the practicality of the absolute value concept in handling distance-related problems within the realm of equations and inequalities.

Multiple Choice Questions

1. Which of the following best describes the absolute value of a number?

    (a) The number's sign (positive or negative)

    (b) The number's distance from zero on the number line

    (c) The reciprocal of the number

    (d) The square of the number

2. Which of the following is the correct piecewise definition of the absolute value function $|x|$?

    (a) $|x| = \begin{cases} x, & \text{if } x < 0, \\ -x, & \text{if } x \geq 0 \end{cases}$

    (b) $|x| = \begin{cases} x, & \text{if } x \geq 0, \\ -x, & \text{if } x < 0 \end{cases}$

    (c) $|x| = \begin{cases} -x, & \text{if } x \geq 0, \\ x, & \text{if } x < 0 \end{cases}$

    (d) $|x| = x^2$

3. Which property of the absolute value function shows that $|x| = |-x|$ for any real number x?

(a) Non-negativity

(b) Multiplicative Property

(c) Even Function Property

(d) Triangle Inequality

4. How many solutions does the equation $|2x - 3| = 5$ have?

   (a) One solution

   (b) Two solutions

   (c) No solution

   (d) An infinite number of solutions

5. What is the outcome when solving an equation of the form $|A(x)| = B$ if B is negative?

   (a) There are two solutions

   (b) There is one unique solution

   (c) There are no solutions

   (d) There are infinitely many solutions

6. Which of the following best represents the Triangle Inequality property of absolute value?

   (a) $|a + b| \geq |a| + |b|$

   (b) $|a + b| = |a| + |b|$

   (c) $|a + b| \leq |a| + |b|$

   (d) $|a - b| = ||a| - |b||$

7. For the inequality $|x - 2| < 3$, which compound inequality correctly describes the solution?

   (a) $x - 2 < -3$ or $x - 2 > 3$

   (b) $-3 < x - 2 < 3$

   (c) $x < -1$ or $x > 5$

   (d) $x - 2 \geq -3$ and $x - 2 \leq 3$

Answers:

1. **B: The number's distance from zero on the number line**
   Explanation: Absolute value measures the nonnegative distance of a number from zero, regardless of its direction.

2. **B:** $|x| = \begin{cases} x, & \text{if } x \geq 0, \\ -x, & \text{if } x < 0 \end{cases}$
   Explanation: This is the standard definition of absolute value, ensuring that the output is always nonnegative.

3. **C: Even Function Property**
   Explanation: The property that $|x| = |-x|$ for all real x demonstrates that the absolute value function is symmetric about the y-axis, making it an even function.

4. **B: Two solutions**
   Explanation: The equation $|2x - 3| = 5$ splits into two cases: $2x - 3 = 5$ and $2x - 3 = -5$, resulting in the solutions $x = 4$ and $x = -1$.

5. **C: There are no solutions**
   Explanation: Since the absolute value of any real number is never negative, an equation like $|A(x)| = B$ with $B < 0$ has no solutions.

6. **C:** $|a + b| \leq |a| + |b|$
   Explanation: This is the triangle inequality, which states that the absolute value of a sum is less than or equal to the sum of the absolute values of the individual terms.

7. **B:** $-3 < x - 2 < 3$
   Explanation: The inequality $|x-2| < 3$ means that the difference $x - 2$ lies between -3 and 3. This compound inequality, when solved, leads to the solution set $-1 < x < 5$.

Practice Problems

1. Write the absolute value function as a piecewise function.
$$|x| = \begin{cases} x, & \text{if } x \geq 0, \\ -x, & \text{if } x < 0. \end{cases}$$

2. Solve the equation:
$$|2x - 5| = 9.$$

3. Solve the inequality:
$$|x + 3| < 4.$$

4. Solve the inequality:
$$|3x - 2| \geq 10.$$

5. Simplify the expression using the multiplicative property of absolute value:
$$|(-4)(x+2)|.$$

6. Verify the triangle inequality for $a = 3$ and $b = -7$ by showing that:
$$|a+b| \leq |a| + |b|.$$

Answers

1. **Piecewise Definition of $|x|$:** The absolute value function is defined as:
$$|x| = \begin{cases} x, & \text{if } x \geq 0, \\ -x, & \text{if } x < 0. \end{cases}$$

   Explanation: This definition states that if x is nonnegative ($x \geq 0$), then its absolute value is simply x. If x is negative ($x < 0$), the function takes the negative of x (which is positive) so that the output is always nonnegative.

2. **Solving the Equation** $|2x - 5| = 9$: Since absolute value equations $|A| = B$ (with $B \geq 0$) have two cases, we set:

   Case 1:
   $$2x - 5 = 9.$$
   Adding 5 to both sides:
   $$2x = 14 \implies x = 7.$$

   Case 2:
   $$2x - 5 = -9.$$
   Adding 5 to both sides:
   $$2x = -4 \implies x = -2.$$

   Explanation: We consider both the positive and negative scenarios because the absolute value of a number is its distance from zero. Thus, the equation has two solutions: $x = 7$ and $x = -2$.

3. **Solving the Inequality** $|x+3| < 4$: The inequality $|A| < B$ (with $B > 0$) can be rewritten as a compound inequality:
   $$-4 < x + 3 < 4.$$
   Subtract 3 from all parts:
   $$-7 < x < 1.$$

   Explanation: This shows that x must lie between -7 and 1 in order for the absolute value of $x + 3$ to be less than 4.

4. **Solving the Inequality** $|3x - 2| \geq 10$: For the inequality $|A| \geq B$, we split it into two cases:

   Case 1:
   $$3x - 2 \geq 10.$$
   Add 2 to both sides:
   $$3x \geq 12 \implies x \geq 4.$$

   Case 2:
   $$3x - 2 \leq -10.$$

Add 2 to both sides:
$$3x \leq -8 \implies x \leq -\frac{8}{3}.$$

Explanation: The inequality indicates that x must satisfy either $x \geq 4$ or $x \leq -\frac{8}{3}$ since the absolute value of the expression will be at least 10 in those regions.

5. **Simplifying $|(-4)(x+2)|$ Using the Multiplicative Property:** The property $|ab| = |a| \cdot |b|$ allows us to write:
$$|(-4)(x+2)| = |-4| \cdot |x+2|.$$

Since $|-4| = 4$, the expression simplifies to:
$$4|x+2|.$$

Explanation: This property highlights that the absolute value of a product is the product of the absolute values, which is useful in simplifying expressions.

6. **Verifying the Triangle Inequality for $a = 3$ and $b = -7$:** First, calculate the absolute value of the sum:
$$|a+b| = |3+(-7)| = |-4| = 4.$$

Then calculate the sum of the absolute values:
$$|a| + |b| = |3| + |-7| = 3 + 7 = 10.$$

Since:
$$4 \leq 10,$$
the triangle inequality
$$|a+b| \leq |a| + |b|$$
is verified.

Explanation: The triangle inequality states that the absolute value of a sum is less than or equal to the sum of the absolute values. In this example, it confirms that the direct distance (4) is less than the total distance traveled when considering individual distances (10).

Chapter 33

Understanding Exponents

Definition of Exponents

An exponent is a notation that indicates how many times a number, known as the base, is multiplied by itself. In mathematical terms, the expression
$$a^n$$
represents the product of the base a multiplied by itself n times. This concept is defined by the equation
$$a^n = \underbrace{a \times a \times \cdots \times a}_{n \text{ times}},$$
where a is any real number and n is a positive integer. This expression provides a concise method for writing repeated multiplication.

Exponents as Repeated Multiplication

The exponent notation allows lengthy multiplications to be expressed compactly. For instance, the expression
$$3^4$$
is a shorthand way to denote the multiplication
$$3^4 = 3 \times 3 \times 3 \times 3.$$

This method efficiently captures the concept of performing the same multiplication multiple times. Repeated multiplication, which may otherwise require extensive writing, is represented in a single, manageable expression using exponents.

Notation and Terminology

The notation a^n consists of two distinct elements: the base a and the exponent n. The base is the number being repeatedly multiplied, while the exponent specifies the number of times the base appears as a factor. Certain special cases are noteworthy:

- When the exponent is 1, the expression simplifies to
$$a^1 = a,$$
indicating that the base is used just once.

- When the exponent is 0 and provided that $a \neq 0$, the expression is defined as
$$a^0 = 1.$$

This outcome follows from the properties of multiplication and the concept of the multiplicative identity.

Illustrative Examples

Several numerical examples illustrate the principle of exponentiation. For example, consider the expression
$$2^5.$$
This is equivalent to
$$2^5 = 2 \times 2 \times 2 \times 2 \times 2.$$
Another illustration is given by
$$5^3,$$
which represents
$$5^3 = 5 \times 5 \times 5.$$
These examples emphasize how exponent notation simplifies the expression of multiple identical factors and allows for efficient calculation and understanding of multiplication repeated several times.

Special Cases and Fundamental Rules

Understanding certain cases of exponents is essential for further exploration in mathematics. When the base is raised to the power of 1, the expression simply equals the base. If the exponent is 0, the result is defined as 1 regardless of the base (with the exception of zero as the base). Moreover, when the base is 1, the expression

$$1^n$$

always evaluates to 1 for any positive integer exponent n. These special cases form the foundation for more advanced work with exponents and promote an understanding of the inherent structure of exponential expressions.

Applications in Mathematical Expressions

Exponentiation plays a vital role in various mathematical topics by providing a method for expressing very large or very small numbers succinctly. Expressing repeated products in compact form enables the use of algebraic rules for simplification and manipulation. The notation a^n serves as a common language in algebra, where it facilitates the handling of equations and the study of patterns across different mathematical contexts. This foundational notation also contributes to the comprehension of more advanced topics, where exponentiation is repeated in complex expressions.

Multiple Choice Questions

1. What does the expression a^n represent?

    (a) a added to itself n times.
    (b) a multiplied by n.
    (c) a multiplied by itself n times.
    (d) a divided by n.

2. How is the expression 3^4 best interpreted?

    (a) $3 + 3 + 3 + 3$
    (b) 3×4
    (c) $3 \times 3 \times 3 \times 3$

(d) $4 \times 4 \times 4$

3. Which statement best describes the roles of the base and the exponent in a^n?

    (a) The base a is added n times.
    (b) The exponent n indicates the number of times the base a is multiplied by itself.
    (c) The base a is divided by the exponent n.
    (d) The exponent n represents how many times the digit a appears in a number.

4. What is the value of 2^5?

    (a) 10
    (b) 16
    (c) 32
    (d) 64

5. According to the fundamental rules of exponents, what is a^0 (assuming $a \neq 0$)?

    (a) 0
    (b) a
    (c) 1
    (d) Undefined

6. For any positive integer n, what is the value of 1^n?

    (a) $1^n = n$
    (b) $1^n = n \times 1$
    (c) $1^n = 1$
    (d) $1^n = 0$

7. How does exponent notation help simplify mathematical expressions?

    (a) It converts repeated addition into multiplication.
    (b) It replaces multiplication with division.
    (c) It rewrites repeated multiplication in a compact form.
    (d) It changes subtraction into multiplication.

Answers:

1. **C: a multiplied by itself n times.**
   Explanation: The notation a^n is used to represent that the base a appears as a factor n times in a multiplication.

2. **C: $3 \times 3 \times 3 \times 3$.**
   Explanation: 3^4 is a concise way to show that 3 is multiplied by itself four times.

3. **B: The exponent n indicates the number of times the base a is multiplied by itself.**
   Explanation: In the expression a^n, a is the base number and the exponent n tells us how many copies of a are multiplied together.

4. **C: 32.**
   Explanation: Calculating 2^5 means multiplying 2 by itself five times: $2 \times 2 \times 2 \times 2 \times 2 = 32$.

5. **C: 1.**
   Explanation: One of the fundamental rules of exponents is that for any nonzero number a, $a^0 = 1$.

6. **C: $1^n = 1$.**
   Explanation: Multiplying 1 by itself any number of times always results in 1, since 1 is the multiplicative identity.

7. **C: It rewrites repeated multiplication in a compact form.**
   Explanation: Exponent notation is an efficient way to represent long strings of multiplications by writing the base and its repetition count as an exponent.

Practice Problems

1. Express the exponent expression:
$$4^3$$
as repeated multiplication.

2. Evaluate the expression:
$$2^5$$
and explain each step of your calculation.

3. Evaluate and explain the expression:
$$3^1$$
describing what the exponent signifies.

4. Evaluate and explain the special case:
$$7^0$$
addressing why any nonzero number raised to the zero power equals 1.

5. In the expression:
$$6^2$$
identify the base and the exponent, and explain their roles.

6. Write the expanded form of:
$$5^4$$
using the definition of exponents, and then compute its value.

Answers

1. **Expressing 4^3 as Repeated Multiplication:**
   Solution: By the definition of exponents, the expression
   $$4^3$$
   represents the base 4 multiplied by itself 3 times. Thus, we can write
   $$4^3 = 4 \times 4 \times 4.$$
   This is the repeated multiplication form of the exponent expression.

2. **Evaluating 2^5:**
   Solution: The expression
   $$2^5$$
   means multiplying 2 by itself 5 times:
   $$2^5 = 2 \times 2 \times 2 \times 2 \times 2.$$
   Step-by-step:
   - First, compute $2 \times 2 = 4. Next, 4 \times 2 = 8.$
   - Then, $8 \times 2 = 16. Finally, 16 \times 2 = 32.$ Therefore,
   $$2^5 = 32.$$

3. **Evaluating 3^1:**
   Solution: According to the definition of exponents, when the exponent is 1 the expression simply equals the base. Hence,
   $$3^1 = 3.$$
   This means that multiplying the base 3 by itself only once results in 3.

4. **Evaluating 7^0:**
   Solution: For any nonzero number, raising it to the power of 0 is defined to equal 1. Therefore,
   $$7^0 = 1.$$
   This rule is based on the properties of exponents and serves as a fundamental rule in mathematics.

5. **Identifying the Base and Exponent in 6^2:**
   Solution: In the expression
   $$6^2,$$
   the number 6 is the **base** since it is the number being multiplied, and the number 2 is the **exponent** since it indicates that 6 is multiplied by itself 2 times. In expanded form, this means:
   $$6^2 = 6 \times 6.$$

6. **Writing the Expanded Form and Evaluating 5^4:**
   Solution: According to the definition of exponents, the expression
   $$5^4$$
   represents 5 multiplied by itself 4 times:
   $$5^4 = 5 \times 5 \times 5 \times 5.$$
   Now, compute the product step-by-step:
   - First, $5 \times 5 = 25.$ Next, $25 \times 5 = 125.$
- Finally, $125 \times 5 = 625.$ Therefore,
   $$5^4 = 625.$$

Chapter 34

Properties of Exponents

The Product Rule for Exponents

The product rule is a fundamental property that applies when multiplying two exponential expressions with the same base. An expression such as a^m represents the base a multiplied by itself m times, and another expression a^n represents a multiplied by itself n times. When these expressions are multiplied, the factors combine into one sequence of repeated multiplications of a. This operation is expressed mathematically as

$$a^m \cdot a^n = a^{m+n}.$$

The rationale behind this rule is that the total number of times the base a appears as a factor in the product is the sum of its appearances in each expression. For example, in the multiplication $2^3 \cdot 2^4$, the base 2 is used 3 times in the first term and 4 times in the second term, resulting in a total of $3+4=7$ factors of 2. Thus, the product simplifies to 2^7.

The Quotient Rule for Exponents

When dividing two exponential expressions with the same nonzero base, the quotient rule provides a method to simplify the expression by subtracting the exponent in the denominator from the exponent

in the numerator. Let a^m be the numerator and a^n be the denominator, with the restriction that a is not equal to zero. The rule is stated as
$$\frac{a^m}{a^n} = a^{m-n}.$$
This rule works because a^m is a sequence of m multiplications of a and a^n is a sequence of n multiplications that are common to both the numerator and the denominator. The division cancels out n factors of a from the numerator, leaving $m - n$ factors of a. For example, in the expression $\frac{5^6}{5^2}$, the division cancels 2 factors from both the numerator and the denominator, leaving $5^{6-2} = 5^4$.

The Power Rule for Exponents

The power rule applies when an exponential expression is itself raised to an additional power. In an expression of the form $(a^m)^n$, the base a is multiplied by itself m times, and this result is then used as the base for further exponentiation n times. Since each application of the inner exponent m contributes factors of a, the overall number of times a appears is the product of the two exponents. This property is expressed as
$$(a^m)^n = a^{m \cdot n}.$$
For instance, in the expression $(3^2)^3$, the base 3 is first squared to form 3^2 and then raised to the third power, resulting in $3^{2 \cdot 3} = 3^6$. This rule simplifies the process of raising an exponent to another exponent by converting nested exponentiation into a single exponentiation with a product of the original exponents.

These properties—the product rule, the quotient rule, and the power rule—form the basis for simplifying a variety of expressions that involve exponents. They allow for the consolidation and reduction of complex exponential expressions into simpler forms by operating on the exponents themselves.

Multiple Choice Questions

1. Which exponent rule tells us that when multiplying two expressions with the same base, we add their exponents?

    (a) Product Rule

(b) Quotient Rule

(c) Power Rule

(d) Zero Exponent Rule

2. Simplify the expression: $2^3 \cdot 2^4$.

    (a) 2^{12}

    (b) 2^7

    (c) 2^4

    (d) 2^1

3. Which rule is applied when dividing exponential expressions with the same nonzero base?

    (a) Product Rule

    (b) Quotient Rule

    (c) Power Rule

    (d) Subtraction Rule

4. Simplify the expression: $(3^2)^3$.

    (a) 3^5

    (b) 3^6

    (c) 3^9

    (d) 3^{2+3}

5. Simplify the quotient: $\frac{5^6}{5^2}$.

    (a) 5^4

    (b) 5^8

    (c) 5^3

    (d) 5^{6-2}

6. Which exponent property is demonstrated by the equation $(a^m)^n = a^{m \cdot n}$?

    (a) Product Rule

    (b) Quotient Rule

    (c) Power Rule

    (d) Distributive Property

7. Evaluate and simplify: $\frac{10^9}{10^4}$.

    (a) 10^5
    (b) 10^{13}
    (c) 10^4
    (d) 10^{9-4+2}

Answers:

1. **A: Product Rule**
   The product rule states that when multiplying two expressions with the same base, you add their exponents: $a^m \cdot a^n = a^{m+n}$.

2. **B: 2^7**
   Applying the product rule, $2^3 \cdot 2^4 = 2^{3+4} = 2^7$.

3. **B: Quotient Rule**
   The quotient rule governs division with the same base, stating that $\frac{a^m}{a^n} = a^{m-n}$ (with $a \neq 0$).

4. **B: 3^6**
   Using the power rule, $(3^2)^3 = 3^{2 \cdot 3} = 3^6$.

5. **A: 5^4**
   By the quotient rule, $\frac{5^6}{5^2} = 5^{6-2} = 5^4$.

6. **C: Power Rule**
   The equation $(a^m)^n = a^{m \cdot n}$ is an example of the power rule, where an exponent is raised to another exponent.

7. **A: 10^5**
   Applying the quotient rule here gives $\frac{10^9}{10^4} = 10^{9-4} = 10^5$.

Practice Problems

1. Simplify the following expression using the product rule for exponents:
$$2^3 \cdot 2^5$$

2. Simplify the following expression using the quotient rule for exponents:
$$\frac{7^6}{7^2}$$

3. Simplify the following expression using the power rule for exponents:
$$\left(3^2\right)^4$$

4. Simplify the expression by applying a combination of the product, power, and quotient rules:
$$\frac{\left(2^3 \cdot 2^4\right)^2}{2^5}$$

5. Simplify the expression and express your answer using only positive exponents:
$$\frac{5^3}{5^5}$$

6. Simplify the following expression by using the quotient rule followed by the power rule:
$$\left(\frac{3^4}{3^2}\right)^3$$

Answers

1. For the expression
$$2^3 \cdot 2^5,$$
we use the product rule for exponents, which states that when multiplying exponential expressions with the same base, we add the exponents. Thus,
$$2^3 \cdot 2^5 = 2^{3+5} = 2^8.$$
This tells us that the base 2 appears a total of 8 times.

2. For the expression
$$\frac{7^6}{7^2},$$
we apply the quotient rule for exponents. This rule states that when dividing two exponential expressions with the same base, we subtract the exponent in the denominator from the exponent in the numerator. Therefore,
$$\frac{7^6}{7^2} = 7^{6-2} = 7^4.$$

Here, the common factors of 7 cancel out, leaving 7 raised to the fourth power.

3. For the expression
$$\left(3^2\right)^4,$$
we use the power rule for exponents. This rule tells us that when an exponential expression is raised to another power, the exponents multiply. Thus,
$$\left(3^2\right)^4 = 3^{2 \cdot 4} = 3^8.$$

This means that the base 3 is multiplied by itself a total of 8 times.

4. To simplify
$$\frac{\left(2^3 \cdot 2^4\right)^2}{2^5},$$
we first use the product rule inside the parentheses:
$$2^3 \cdot 2^4 = 2^{3+4} = 2^7.$$

Next, we apply the power rule by raising the result to the second power:
$$\left(2^7\right)^2 = 2^{7 \cdot 2} = 2^{14}.$$

Finally, we use the quotient rule to divide by 2^5:
$$\frac{2^{14}}{2^5} = 2^{14-5} = 2^9.$$

So, the simplified expression is 2^9.

5. For the expression
$$\frac{5^3}{5^5},$$
we again use the quotient rule for exponents:
$$\frac{5^3}{5^5} = 5^{3-5} = 5^{-2}.$$
To express the answer with only positive exponents, we recall that a negative exponent indicates the reciprocal:
$$5^{-2} = \frac{1}{5^2}.$$
Thus, the simplified expression with a positive exponent is $\frac{1}{5^2}$.

6. For the expression
$$\left(\frac{3^4}{3^2}\right)^3,$$
we begin by applying the quotient rule within the parentheses:
$$\frac{3^4}{3^2} = 3^{4-2} = 3^2.$$
Then, we use the power rule to raise the result to the third power:
$$\left(3^2\right)^3 = 3^{2 \cdot 3} = 3^6.$$
Therefore, the simplified expression is 3^6.

Chapter 35

Exponents with Negative and Zero Powers

Zero Exponents

When working with exponents, the case in which an exponent is zero requires special attention in order to maintain the consistency of the laws of exponents. For any nonzero number a and any positive integer n, the expression

$$\frac{a^n}{a^n} = a^{n-n} = a^0$$

must equal 1 because the quotient of any nonzero number by itself is 1. This definition is chosen so that the rule remains valid for all integer exponents. In this way, the concept of a zero exponent aligns with the idea that an exponent indicates repeated multiplication, and when there are no factors present in the multiplication, the result is 1. For example, with a base of 5, the expression

$$5^0 = 1,$$

and with a base of -3,

$$(-3)^0 = 1,$$

provided that the base is not zero. This special case assures that exponential expressions continue to behave predictably when the exponents are manipulated according to the established rules.

Negative Exponents

Negative exponents extend the rules of exponentiation by expressing reciprocals. For any nonzero number a and any positive integer n, the definition

$$a^{-n} = \frac{1}{a^n}$$

ensures that the rules governing exponent operations remain applicable even when the exponent is negative. This formulation is not arbitrary; rather, it derives naturally from the requirement that the product of exponential expressions must adhere to the rule

$$a^n \cdot a^{-n} = a^{n-n} = a^0,$$

which, as established, equals 1. For instance, with a base of 2,

$$2^{-3} = \frac{1}{2^3} = \frac{1}{8},$$

and with a base of 10,

$$10^{-2} = \frac{1}{10^2} = \frac{1}{100}.$$

This reciprocal relationship not only simplifies the process of rewriting expressions in a more manageable form but also plays a crucial role in many computational problems by allowing expressions to be reformed in a uniform manner.

Applications in Computations

The rules for zero and negative exponents contribute significantly to the simplification of algebraic expressions and facilitate efficient computations. In many cases, expressions that contain a mixture of positive, zero, and negative exponents can be manipulated using the standard rules of multiplication and division. For example, when multiplying powers of the same base, the process

$$a^n \cdot a^{-n} = a^{n-n} = a^0 = 1$$

demonstrates a cancellation effect that can streamline complex expressions in algebraic equations.

Consider the expression
$$\frac{3^5 \cdot 3^{-2}}{3^3}.$$

By applying the product rule in the numerator,
$$3^5 \cdot 3^{-2} = 3^{5-2} = 3^3,$$
and then using the quotient rule,
$$\frac{3^3}{3^3} = 3^{3-3} = 3^0 = 1,$$
the original expression simplifies neatly to 1. Such simplifications are essential in problem solving because they reduce the complexity of an expression and allow for the clear identification of underlying patterns.

In many computations, rewriting expressions with negative exponents into their reciprocal forms leads to a consistent use of only positive exponents. This practice frequently makes further operations, such as addition or subtraction of similar terms, more straightforward. For instance, an expression like
$$4^{-2} \cdot 4^5$$
can be rearranged utilizing the product rule to give
$$4^{5-2} = 4^3,$$
which emphasizes the utility of the negative exponent rule in obtaining a simplified result.

Attention to the meaning of zero and negative exponents in computations solidifies a deeper understanding of exponentiation and ensures that all properties of exponents remain intact during successive algebraic manipulations.

Multiple Choice Questions

1. Which of the following statements is true about zero exponents?

    (a) For any nonzero number a, $a^0 = 0$.

(b) For any nonzero number a, $a^0 = 1$.

(c) For any number a, $a^0 = a$.

(d) For any nonzero number a, a^0 is undefined.

2. What does the expression 2^{-3} equal?

   (a) $\frac{1}{8}$

   (b) -8

   (c) 8

   (d) $-\frac{1}{8}$

3. Simplify the expression:
$$\frac{3^5 \cdot 3^{-2}}{3^3}.$$

   (a) 3^0

   (b) 3^2

   (c) 3^{-3}

   (d) 3^5

4. The equality
$$a^n \cdot a^{-n} = a^0$$
   is an example of which exponent rule?

   (a) The zero property of multiplication

   (b) The reciprocal property

   (c) The product rule for exponents

   (d) The division rule for exponents

5. Why is it useful to rewrite expressions with negative exponents into their reciprocal forms?

   (a) It makes the expression more complicated.

   (b) It allows us to work with only positive exponents, simplifying computations.

   (c) It automatically changes the base of the expression.

   (d) It is only useful when dividing expressions.

6. For any nonzero number a and positive integer n, the expression a^{-n} is defined as:

(a) a^n

(b) $\frac{1}{a^n}$

(c) $-a^n$

(d) $-\frac{1}{a^n}$

7. Simplify the expression:
$$5^{-2} \cdot 5^7.$$

(a) 5^9

(b) 5^5

(c) 5^{-9}

(d) 5^{-5}

Answers:

1. **B: For any nonzero number a, $a^0 = 1$**
   Explanation: According to the zero exponent rule, for any nonzero base a, dividing a^n by itself (i.e., a^n/a^n) gives $a^{n-n} = a^0$, which must equal 1.

2. **A: $\frac{1}{8}$**
   Explanation: Negative exponents indicate reciprocals. Thus, 2^{-3} is defined as $\frac{1}{2^3}$, and since $2^3 = 8$, the expression equals $\frac{1}{8}$.

3. **A: 3^0**
   Explanation: First, apply the product rule: $3^5 \cdot 3^{-2} = 3^{5-2} = 3^3$. Then, applying the quotient rule with the denominator 3^3:
   $$\frac{3^3}{3^3} = 3^{3-3} = 3^0,$$
   which equals 1.

4. **C: The product rule for exponents**
   Explanation: The product rule states that when multiplying two powers of the same base, you add the exponents. Here, adding n and $-n$ gives $a^{n+(-n)} = a^0$.

5. **B: It allows us to work with only positive exponents, simplifying computations**
   Explanation: Rewriting expressions with negative exponents

into their reciprocal form converts them into expressions using only positive exponents. This makes further arithmetic or algebraic manipulation, such as addition and subtraction of like terms, much more straightforward.

6. **B:** $\frac{1}{a^n}$
   Explanation: By definition, a negative exponent represents the reciprocal of the corresponding positive exponent. Therefore, $a^{-n} = \frac{1}{a^n}$.

7. **B:** 5^5
   Explanation: When multiplying powers with the same base, add the exponents: $-2 + 7 = 5$. Thus, $5^{-2} \cdot 5^7 = 5^5$, which is the simplified form.

Practice Problems

1. Evaluate the following expressions:
$$7^0 \quad \text{and} \quad (-4)^0.$$

2. Evaluate and rewrite the following using their fractional forms:
$$2^{-3} \quad \text{and} \quad 5^{-2}.$$

3. Simplify the expression using the laws of exponents:
$$\frac{3^4 \cdot 3^{-2}}{3^3}.$$

4. Simplify the following expression:
$$4^{-1} \cdot 4^5.$$

5. Simplify the expression:
$$\frac{6^{-2}}{6^{-5}}.$$

6. Simplify and evaluate the expression:
$$\frac{2^3 \cdot 2^{-5}}{2^{-2}}.$$

Answers

1. Evaluate the expressions:
$$7^0 \quad \text{and} \quad (-4)^0.$$

   Solution:
   By definition, any nonzero number raised to the zero power is 1. This is because for any nonzero number a,
   $$a^0 = \frac{a^n}{a^n} = 1 \quad \text{(for any positive integer } n\text{)}.$$
   Thus,
   $$7^0 = 1 \quad \text{and} \quad (-4)^0 = 1.$$
   This consistent rule ensures that the laws of exponents remain valid even when the exponent is zero.

2. Evaluate and rewrite using fractional forms:
$$2^{-3} \quad \text{and} \quad 5^{-2}.$$

   Solution:
   The negative exponent rule states that for any nonzero number a and positive integer n,
   $$a^{-n} = \frac{1}{a^n}.$$

Therefore,
$$2^{-3} = \frac{1}{2^3} = \frac{1}{8},$$
and
$$5^{-2} = \frac{1}{5^2} = \frac{1}{25}.$$

This reciprocal relationship allows us to express negative exponents as fractions.

3. Simplify the expression:
$$\frac{3^4 \cdot 3^{-2}}{3^3}.$$

Solution:
First, apply the product rule of exponents in the numerator:
$$3^4 \cdot 3^{-2} = 3^{4+(-2)} = 3^2.$$

Next, use the quotient rule:
$$\frac{3^2}{3^3} = 3^{2-3} = 3^{-1}.$$

Finally, rewrite 3^{-1} as a fraction:
$$3^{-1} = \frac{1}{3}.$$

Hence, the simplified expression is $\frac{1}{3}$.

4. Simplify the expression:
$$4^{-1} \cdot 4^5.$$

Solution:
Using the product rule for exponents:
$$4^{-1} \cdot 4^5 = 4^{-1+5} = 4^4.$$

Evaluating 4^4:
$$4^4 = 4 \times 4 \times 4 \times 4 = 256.$$

Therefore, the simplified result is 256.

5. Simplify the expression:
$$\frac{6^{-2}}{6^{-5}}.$$

Solution:
Apply the quotient rule $a^m/a^n = a^{m-n}$:
$$\frac{6^{-2}}{6^{-5}} = 6^{-2-(-5)} = 6^{-2+5} = 6^3.$$

Calculating 6^3:
$$6^3 = 6 \times 6 \times 6 = 216.$$

Thus, the simplified expression evaluates to 216.

6. Simplify and evaluate the expression:
$$\frac{2^3 \cdot 2^{-5}}{2^{-2}}.$$

Solution:
Start by applying the product rule in the numerator:
$$2^3 \cdot 2^{-5} = 2^{3+(-5)} = 2^{-2}.$$

Then, apply the quotient rule:
$$\frac{2^{-2}}{2^{-2}} = 2^{-2-(-2)} = 2^0.$$

Since any nonzero number raised to the zero power equals 1,
$$2^0 = 1.$$

Therefore, the final simplified result is 1.

Chapter 36

Introduction to Radicals

Definition and Notation of Radicals

Radicals provide a compact notation to express the inverse operation of exponentiation. The symbol $\sqrt{}$ is used primarily to denote the square root of a nonnegative number. In the expression $\sqrt{x}$, the number x is known as the radicand, and the radical symbol indicates the operation of finding the unique nonnegative number which, when squared, equals x. This definition is based on the convention that squaring a number and then taking the square root returns the original nonnegative number. In mathematical terms, if $y = \sqrt{x}$, then the equality

$$y^2 = x$$

holds for every nonnegative x.

Square Roots as the Inverse of Exponentiation

The square root operation is intrinsically linked to the process of exponentiation. When a number is raised to the second power (squared), the square root operation reverses this process. This inverse relationship is captured by the equivalence

$$\sqrt{x} = x^{\frac{1}{2}},$$

which expresses the square root as a fractional exponent. Here, raising a number to the power $\frac{1}{2}$ is defined to uniquely reverse the squaring of its positive square root. For example, if a number a is squared to yield a^2, then the square root of a^2 is given as

$$\sqrt{a^2} = a,$$

provided that a is nonnegative. This property demonstrates that the process of taking square roots undoes the effect of squaring and maintains the consistency of exponentiation rules.

Properties of Square Roots

The algebraic properties of square roots facilitate the simplification and manipulation of expressions involving radicals. One key property is the product rule for radicals, which states that for any two nonnegative numbers a and b, the square root of their product equals the product of their individual square roots:

$$\sqrt{ab} = \sqrt{a}\sqrt{b}.$$

This property follows naturally from the rules of exponentiation because writing a and b in their exponential forms as $a^{\frac{1}{2}}$ and $b^{\frac{1}{2}}$ respectively, yields

$$a^{\frac{1}{2}} \cdot b^{\frac{1}{2}} = (ab)^{\frac{1}{2}}.$$

Another important concept is the notion that radicals express the inverse of the exponentiation process. For any nonnegative number x and any positive integer n, the nth root of x is defined as the number that, when raised to the power n, equals x. In the special case when $n = 2$, the square root notation is used. This inverse relationship reinforces the idea that exponentiation and taking roots are mutually reversible operations. Written in exponential notation, the square root is expressed as

$$x^{\frac{1}{2}},$$

which clearly distinguishes the operation as the inverse of squaring.

Moreover, the properties of radicals and exponents extend to allow expressions with fractional exponents to be manipulated using familiar algebraic rules. If a number is raised to a rational exponent, such as $\frac{m}{n}$, the expression can be interpreted as taking the nth root of the number raised to the mth power:

$$x^{\frac{m}{n}} = \left(\sqrt[n]{x}\right)^m.$$

In the context of square roots, where $n = 2$, this formulation becomes
$$x^{\frac{m}{2}} = \left(\sqrt{x}\right)^m,$$
which illustrates the seamless connection between radical notation and exponentiation.

Interpretation of Square Roots in Computation

The concept of square roots as the inverse of exponentiation not only provides a theoretical framework but also serves practical computational purposes. Rewriting square roots using fractional exponents enables the application of exponent rules for multiplication, division, and power operations. For instance, the expression
$$\sqrt{a} \cdot \sqrt{a} = a^{\frac{1}{2}} \cdot a^{\frac{1}{2}}$$
can be simplified by adding the exponents:
$$a^{\frac{1}{2} + \frac{1}{2}} = a^1 = a.$$

This technique offers a consistent method for simplifying more intricate expressions and demonstrates the power of understanding radicals in terms of exponentiation. Such clarity in interpretation ensures that expressions involving radicals are handled in a mathematically robust and coherent manner.

Multiple Choice Questions

1. Which of the following statements best defines the radical symbol $\sqrt{}$ when applied to a nonnegative number x?

    (a) It denotes the cube root of x.

    (b) It denotes the negative square root of x.

    (c) It denotes the unique nonnegative square root of x.

    (d) It denotes raising x to the power of 2.

2. In the expression $\sqrt{x}$, what is the term used to describe the number x?

    (a) Exponent

(b) Coefficient

   (c) Radicand

   (d) Base

3. How can the square root of x be expressed using exponents?

   (a) x^2

   (b) $x^{-\frac{1}{2}}$

   (c) $x^{\frac{1}{2}}$

   (d) $\frac{1}{x^{\frac{1}{2}}}$

4. Which property of radicals correctly shows the relationship between the square roots of two nonnegative numbers a and b?

   (a) $\sqrt{ab} = \sqrt{a} + \sqrt{b}$
   (b) $\sqrt{ab} = \sqrt{a} - \sqrt{b}$
   (c) $\sqrt{ab} = \sqrt{a}\,\sqrt{b}$
   (d) $\sqrt{ab} = a\,\sqrt{b}$

5. What is the radical (square root) notation equivalent to $x^{\frac{1}{2}}$?

   (a) $\sqrt{x}$

   (b) $x\sqrt{}$

   (c) $\frac{1}{\sqrt{x}}$

   (d) x^2

6. For any nonnegative number a, which of the following correctly simplifies the expression $\sqrt{a} \cdot \sqrt{a}$?

   (a) $\sqrt{a^2}$

   (b) a

   (c) Both A and B

   (d) a^2

7. Which statement best describes the relationship between squaring a number and taking its square root?

   (a) They are two identical operations that always yield the same result.

(b) Squaring a number always yields the original number when reversed by square rooting.

(c) Taking the square root reverses the process of squaring, yielding the original nonnegative number.

(d) There is no direct relationship between squaring and taking square roots.

Answers:

1. **C:** It denotes the unique nonnegative square root of x.
   Explanation: By convention, the radical symbol $\sqrt{}$ represents the principal (nonnegative) square root of a nonnegative number.

2. **C:** Radicand
   Explanation: In the notation $\sqrt{x}$, the number x under the radical is called the radicand.

3. **C:** $x^{\frac{1}{2}}$
   Explanation: The square root of x can be expressed as $x^{\frac{1}{2}}$, which directly follows from the rules of exponents.

4. **C:** $\sqrt{ab} = \sqrt{a}\sqrt{b}$
   Explanation: This is the product rule for radicals; for any nonnegative numbers a and b, the square root of their product is equal to the product of their individual square roots.

5. **A:** $\sqrt{x}$
   Explanation: The expression $x^{\frac{1}{2}}$ is another way of writing the square root of x.

6. **C:** Both A and B
   Explanation: Multiplying $\sqrt{a}$ by itself gives $\sqrt{a} \cdot \sqrt{a} = \sqrt{a^2}$, and since a is nonnegative, $\sqrt{a^2} = a$.

7. **C:** Taking the square root reverses the process of squaring, yielding the original nonnegative number.
   Explanation: Squaring a number and then taking its square root returns the initial nonnegative number, highlighting the inverse nature of these operations.

Practice Problems

1. What does the radical symbol $\sqrt{}$ represent, and what is the radicand? Provide an example using a nonnegative number.

2. Rewrite the following square root expressions using fractional exponents:
$$\sqrt{b} \quad \text{and} \quad \sqrt{b^4}.$$

3. Use the product rule for radicals to simplify the expression:
$$\sqrt{3} \cdot \sqrt{12}.$$

4. Explain how the square root operation acts as the inverse of squaring. Illustrate your explanation with a numerical example.

5. Express $16^{\frac{3}{2}}$ in radical form and compute its value.

6. Simplify the following expression using properties of radicals:

$$\sqrt{50} \cdot \sqrt{8}.$$

Answers

1. **Solution:** The radical symbol $\sqrt{}$ denotes the square root, which is defined as the nonnegative number whose square is the radicand (the number under the radical sign). For example, in the expression $\sqrt{9}$, the radicand is 9. Since $3^2 = 9$ and we choose the nonnegative value, $\sqrt{9} = 3$. This example shows that the square root operation reverses squaring, provided the original number is nonnegative.

2. **Solution:** To rewrite a square root using fractional exponents, recall that the square root of a number is equivalent to raising it to the power of $\frac{1}{2}$. Therefore,
$$\sqrt{b} = b^{\frac{1}{2}},$$
and for the second expression,
$$\sqrt{b^4} = \left(b^4\right)^{\frac{1}{2}} = b^{4 \cdot \frac{1}{2}} = b^2.$$
These conversions illustrate the relationship between radical notation and fractional exponents.

3. **Solution:** The product rule for radicals states that for any two nonnegative numbers a and b,
$$\sqrt{a} \cdot \sqrt{b} = \sqrt{ab}.$$
Applying this to the expression at hand:
$$\sqrt{3} \cdot \sqrt{12} = \sqrt{3 \times 12} = \sqrt{36}.$$
Since $\sqrt{36} = 6$, the simplified expression is 6.

4. **Solution:** The square root operation is the inverse of squaring because it "undoes" the squaring process. If you take a nonnegative number x, square it to obtain x^2, and then take the square root of the result, you return to x. For example, if $x = 5$, then:
$$5^2 = 25,$$
and taking the square root,
$$\sqrt{25} = 5.$$
This demonstrates that the operations of squaring and taking the square root are inverses (for nonnegative numbers), making the square root a reversing process of squaring.

5. **Solution:** The expression $16^{\frac{3}{2}}$ can be rewritten in radical form using the rule:
$$x^{\frac{m}{n}} = \left(\sqrt[n]{x}\right)^m.$$
With $x = 16$, $m = 3$, and $n = 2$, we have:
$$16^{\frac{3}{2}} = \left(\sqrt{16}\right)^3.$$
Since $\sqrt{16} = 4$, it follows that:
$$(4)^3 = 64.$$
Therefore, $16^{\frac{3}{2}} = 64$.

6. **Solution:** To simplify $\sqrt{50} \cdot \sqrt{8}$, we first apply the product rule for radicals:
$$\sqrt{50} \cdot \sqrt{8} = \sqrt{50 \times 8} = \sqrt{400}.$$
Recognizing that 400 is a perfect square, we find:
$$\sqrt{400} = 20.$$
Thus, the simplified expression is 20.

Chapter 37

Simplifying Radical Expressions

Techniques for Extracting Perfect Square Factors

Radical expressions can often be written in a simplified form by identifying and extracting perfect square factors from the radicand. An expression of the form
$$\sqrt{n}$$
is considered simplified when no factor of n is a perfect square larger than 1. This process relies on the product property of radicals, which states that for any nonnegative numbers a and b,
$$\sqrt{a \cdot b} = \sqrt{a}\,\sqrt{b}.$$
For instance, consider the expression
$$\sqrt{72}.$$
Note that 72 may be factored as 36×2, where 36 is a perfect square since $6^2 = 36$. Applying the product property yields
$$\sqrt{72} = \sqrt{36 \times 2} = \sqrt{36}\,\sqrt{2} = 6\sqrt{2}.$$
This method involves reorganizing the radicand into a product of a perfect square and another factor, and then simplifying by taking the square root of the perfect square.

An alternative approach utilizes the prime factorization of the radicand. Writing the radicand as a product of its prime factors can reveal pairs of identical factors. Since each pair represents a perfect square, it is possible to extract the factor outside the radical symbol. For example, consider

$$\sqrt{200}.$$

The prime factorization of 200 is

$$2^3 \times 5^2.$$

Grouping identical factors, the expression can be rearranged as

$$\sqrt{2^2 \cdot 5^2 \cdot 2} = \sqrt{(2 \cdot 5)^2 \cdot 2} = \sqrt{100 \cdot 2}.$$

Then, by extracting the perfect square,

$$\sqrt{200} = \sqrt{100}\sqrt{2} = 10\sqrt{2}.$$

This systematic extraction improves clarity and facilitates subsequent operations involving the expression.

Rationalizing Denominators

When a radical occurs in the denominator of a fractional expression, it is often desirable to transform the expression so that the denominator is a rational number. The process of eliminating radicals from the denominator is known as rationalizing the denominator.

For a simple radical in the denominator, consider the fraction

$$\frac{1}{\sqrt{3}}.$$

Multiplying both the numerator and the denominator by $\sqrt{3}$ provides

$$\frac{1}{\sqrt{3}} \times \frac{\sqrt{3}}{\sqrt{3}} = \frac{\sqrt{3}}{(\sqrt{3})^2} = \frac{\sqrt{3}}{3}.$$

This multiplication removes the radical from the denominator because the square of a square root returns the original nonnegative number.

In cases where the denominator is formed by a sum or difference involving a radical, the method involves multiplication by the conjugate. Consider an expression of the form

$$\frac{1}{a + b\sqrt{c}},$$

where a, b, and c are constants with c a positive number that is not a perfect square. The conjugate of the denominator is

$$a - b\sqrt{c}.$$

Multiplying the numerator and the denominator by this conjugate yields

$$\frac{1}{a + b\sqrt{c}} \times \frac{a - b\sqrt{c}}{a - b\sqrt{c}} = \frac{a - b\sqrt{c}}{(a + b\sqrt{c})(a - b\sqrt{c})}.$$

The denominator simplifies via the difference of two squares:

$$(a + b\sqrt{c})(a - b\sqrt{c}) = a^2 - (b\sqrt{c})^2 = a^2 - b^2 c.$$

This results in an expression with a rational denominator, assuming that $a^2 - b^2 c$ is nonzero. The process guarantees that the radical is completely removed from the denominator of the original fraction.

A further example involves a common scenario such as

$$\frac{1}{2 - \sqrt{5}}.$$

Multiplying numerator and denominator by the conjugate,

$$2 + \sqrt{5},$$

yields

$$\frac{1}{2 - \sqrt{5}} \times \frac{2 + \sqrt{5}}{2 + \sqrt{5}} = \frac{2 + \sqrt{5}}{(2)^2 - (\sqrt{5})^2} = \frac{2 + \sqrt{5}}{4 - 5} = \frac{2 + \sqrt{5}}{-1}.$$

The final form,

$$-(2 + \sqrt{5}),$$

exhibits a rational denominator and illustrates the effectiveness of using the conjugate technique.

The techniques described in these sections provide systematic methods for both simplifying radical expressions through extraction of perfect square factors and for rationalizing denominators. These methods enable consistent manipulation and reduction of expressions containing radicals, paving the way for more complex algebraic operations.

Multiple Choice Questions

1. Which property of radicals allows us to write an expression of the form $\sqrt{a \cdot b}$ as $\sqrt{a}\,\sqrt{b}$ (for nonnegative a and b)?

    (a) The quotient property of radicals

    (b) The product property of radicals

    (c) The sum property of radicals

    (d) The difference property of radicals

2. What is the simplified form of $\sqrt{72}$?

    (a) $6\sqrt{2}$

    (b) $2\sqrt{18}$

    (c) $3\sqrt{8}$

    (d) $4\sqrt{3}$

3. By using prime factorization, $\sqrt{200}$ can be simplified. Which of the following correctly shows its simplified form?

    (a) $\sqrt{200} = 10\sqrt{2}$

    (b) $\sqrt{200} = 5\sqrt{2}$

    (c) $\sqrt{200} = 2\sqrt{50}$

    (d) $\sqrt{200} = 4\sqrt{2}$

4. What is the result of rationalizing the denominator of the expression $\frac{1}{\sqrt{3}}$?

    (a) $\frac{\sqrt{3}}{3}$

    (b) $\frac{1}{3\sqrt{3}}$

    (c) $\frac{\sqrt{3}}{2}$

    (d) $\sqrt{3}$

5. To rationalize a denominator of the form $a + b\sqrt{c}$ (with c not a perfect square), which technique should be used?

    (a) Multiply the numerator and denominator by $a - b\sqrt{c}$

    (b) Multiply the numerator and denominator by $\sqrt{a + b\sqrt{c}}$

    (c) Multiply the numerator and denominator by $a + b\sqrt{c}$

(d) Multiply the numerator and denominator by 2

6. Simplify the expression $\frac{1}{2-\sqrt{5}}$ by rationalizing the denominator.

    (a) $-(2+\sqrt{5})$
    (b) $2+\sqrt{5}$
    (c) $\frac{2+\sqrt{5}}{3}$
    (d) $-(2-\sqrt{5})$

7. Which of the following radical expressions is NOT in its simplest form?

    (a) $2\sqrt{5}$
    (b) $3\sqrt{2}$
    (c) $\sqrt{20}$
    (d) $4\sqrt{3}$

Answers:

1. **B: The product property of radicals**
   This property states that $\sqrt{a \cdot b} = \sqrt{a}\sqrt{b}$ for any nonnegative numbers a and b. It is fundamental when breaking apart or combining factors under one radical.

2. **A: $6\sqrt{2}$**
   Since $72 = 36 \times 2$ and $\sqrt{36} = 6$, we can write $\sqrt{72} = \sqrt{36 \times 2} = 6\sqrt{2}$.

3. **A: $\sqrt{200} = 10\sqrt{2}$**
   The prime factorization of 200 is $2^3 \times 5^2$. Rearranging gives $200 = (2^2 \cdot 5^2) \cdot 2 = 100 \cdot 2$, so $\sqrt{200} = \sqrt{100 \cdot 2} = 10\sqrt{2}$.

4. **A: $\frac{\sqrt{3}}{3}$**
   Multiply the expression by $\frac{\sqrt{3}}{\sqrt{3}}$ to obtain $\frac{\sqrt{3}}{(\sqrt{3})^2} = \frac{\sqrt{3}}{3}$, which removes the radical from the denominator.

5. **A: Multiply the numerator and denominator by $a - b\sqrt{c}$**
   Multiplying by the conjugate $a - b\sqrt{c}$ transforms the denominator into a difference of two squares, thereby eliminating the radical and rationalizing the denominator.

6. **A:** $-\left(2+\sqrt{5}\right)$

   Multiplying by the conjugate yields: $\frac{1}{2-\sqrt{5}} \times \frac{2+\sqrt{5}}{2+\sqrt{5}} = \frac{2+\sqrt{5}}{4-5} = \frac{2+\sqrt{5}}{-1} = -(2+\sqrt{5})$.

7. **C:** $\sqrt{20}$

   The expression $\sqrt{20}$ is not in simplest form because 20 can be expressed as 4×5 and $\sqrt{4} = 2$. Thus, $\sqrt{20}$ simplifies to $2\sqrt{5}$, while the other options are already fully simplified.

Practice Problems

1. Simplify the following radical expression by extracting perfect square factors:
$$\sqrt{50}$$

2. Simplify the following radical expression by identifying a perfect square factor:
$$\sqrt{72}$$

3. Simplify the following radical expression using prime factorization:
$$\sqrt{98}$$

4. Rationalize the denominator of the following fraction:
$$\frac{3}{\sqrt{2}}$$

5. Rationalize the denominator of the following expression using the conjugate method:
$$\frac{1}{3 - \sqrt{5}}$$

6. Simplify and rationalize the following expression:
$$\frac{\sqrt{72}}{2\sqrt{8}}$$

Answers

1. **Solution:** To simplify $\sqrt{50}$, first express 50 as a product of a perfect square and another factor:
$$50 = 25 \times 2.$$
Then, the square root can be written as:
$$\sqrt{50} = \sqrt{25 \times 2} = \sqrt{25}\sqrt{2} = 5\sqrt{2}.$$
Therefore, the simplified form is:
$$5\sqrt{2}.$$

2. **Solution:** The number 72 can be factored to find its perfect square:
$$72 = 36 \times 2.$$
Then, applying the product property of radicals:
$$\sqrt{72} = \sqrt{36 \times 2} = \sqrt{36}\sqrt{2} = 6\sqrt{2}.$$
Thus, the simplified expression is:
$$6\sqrt{2}.$$

3. **Solution:** Begin by finding the prime factorization of 98:
$$98 = 2 \times 49 \quad \text{and} \quad 49 = 7^2.$$
Rewrite the radicand as:
$$\sqrt{98} = \sqrt{7^2 \times 2} = \sqrt{7^2}\sqrt{2} = 7\sqrt{2}.$$
Therefore, the expression simplifies to:
$$7\sqrt{2}.$$

4. **Solution:** To rationalize the denominator of $\frac{3}{\sqrt{2}}$, multiply the numerator and denominator by $\sqrt{2}$ so that the denominator becomes rational:
$$\frac{3}{\sqrt{2}} \times \frac{\sqrt{2}}{\sqrt{2}} = \frac{3\sqrt{2}}{(\sqrt{2})^2} = \frac{3\sqrt{2}}{2}.$$
Hence, the rationalized form is:
$$\frac{3\sqrt{2}}{2}.$$

5. **Solution:** For the expression $\frac{1}{3-\sqrt{5}}$, we use the conjugate. The conjugate of $3 - \sqrt{5}$ is $3 + \sqrt{5}$. Multiply both the numerator and the denominator by the conjugate:
$$\frac{1}{3-\sqrt{5}} \times \frac{3+\sqrt{5}}{3+\sqrt{5}} = \frac{3+\sqrt{5}}{(3-\sqrt{5})(3+\sqrt{5})}.$$
The denominator simplifies using the difference of two squares:
$$(3-\sqrt{5})(3+\sqrt{5}) = 3^2 - (\sqrt{5})^2 = 9 - 5 = 4.$$
Thus, the simplified expression is:
$$\frac{3+\sqrt{5}}{4}.$$

6. **Solution:** First, simplify the numerator and the denominator separately.

   For the numerator:
$$\sqrt{72} = \sqrt{36 \times 2} = \sqrt{36}\sqrt{2} = 6\sqrt{2}.$$

For the denominator, note that:
$$\sqrt{8} = \sqrt{4 \times 2} = \sqrt{4}\sqrt{2} = 2\sqrt{2}.$$

So,
$$2\sqrt{8} = 2(2\sqrt{2}) = 4\sqrt{2}.$$

The expression becomes:
$$\frac{6\sqrt{2}}{4\sqrt{2}}.$$

Cancel the common factor of $\sqrt{2}$ (since $\sqrt{2} \neq 0$):
$$\frac{6}{4} = \frac{3}{2}.$$

Therefore, the simplified and rationalized expression is:
$$\frac{3}{2}.$$

Chapter 38

Scientific Notation

Definition and Fundamental Principles

Scientific notation is a method by which numbers, whether extremely large or exceedingly small, are expressed in a compact and standardized form. In this form, a number is represented as the product of a decimal coefficient and a power of 10. The coefficient is a value that lies between 1 (inclusive) and 10 (exclusive), and the exponent on 10 indicates the number of places the decimal point has been shifted. When the exponent is positive, it implies that the original number is large and the decimal point is moved to the left. Conversely, a negative exponent signifies that the original number is small and the decimal point is moved to the right. This notation facilitates easier comparison, computation, and communication of values by reducing long strings of digits into a more manageable format.

Expressing Very Large Numbers in Compact Form

When a large number is written in its standard numerical form, it often contains numerous digits that can be unwieldy to work with. Converting such a number to scientific notation involves repositioning the decimal point so that only one nonzero digit remains to the left of the decimal. For instance, a large whole number can be transformed by counting the number of places the decimal point is moved from its original position at the end of the number to the new position immediately after the first digit. This count becomes the positive exponent of 10. The resulting expression is concise and clearly illustrates the scale of the number. Illustrative examples in mathematical notation are as follows:

$$6\,500\,000 = 6.5 \times 10^6$$

$$9\,200\,000\,000 = 9.2 \times 10^9$$

These expressions demonstrate the systematic procedure in which every large number is reduced to a coefficient containing one significant digit to the left of the decimal point and a corresponding power of 10 that captures the magnitude of the original number.

Expressing Very Small Numbers in Compact Form

Very small numbers, particularly those less than 1, display many zeros when written in standard decimal form. Scientific notation offers a means to express these numbers in a succinct fashion by shifting the decimal point to create a coefficient between 1 and 10. In this scenario, the decimal point is moved to the right, and the number of shifts is recorded as a negative exponent on 10. This negative exponent effectively communicates the position of the decimal point in the original number. A detailed example is shown below:

$$0.00042 = 4.2 \times 10^{-4}$$

In this conversion, the decimal point is moved four places to the right to position the significant digits correctly, and the exponent -4 indicates this shift. The methodology ensures that even the smallest quantities are represented in a manner that emphasizes their essential numerical value while eliminating unnecessary zeros.

Conversion Process and Detailed Examples

The conversion process to scientific notation requires careful attention to the placement of the decimal point and an accurate count of the shifts needed. The procedure can be divided into distinct steps. First, identify the significant digits by locating the first nonzero digit. Second, reposition the decimal point so that it comes immediately after this first nonzero digit. Third, determine the appropriate exponent by counting the number of positions the decimal point has been moved. For numbers greater than or equal to 10, the count is recorded as a positive integer; for numbers less than 1, the count is recorded as a negative integer.

Consider the conversion of a large number:

$$47\,000\,000 = 4.7 \times 10^7$$

In this example, the decimal point originally at the end of the number is moved seven places to the left, resulting in the exponent 7. Similarly, for a small number:

$$0.00089 = 8.9 \times 10^{-4}$$

Here, moving the decimal point four places to the right produces the exponent -4. The clarity provided by these transformations underscores the power of scientific notation as an effective method for rendering numbers both compact and comprehensible, especially when dealing with extreme orders of magnitude.

Multiple Choice Questions

1. Which of the following best describes scientific notation?

    (a) Expressing a number as the sum of a decimal and a power of 10.

    (b) Expressing a number as the product of a decimal coefficient and a power of 10, where the coefficient is at least 1 but less than 10.

    (c) Expressing a number solely as a power of 10.

    (d) Expressing a number as a fraction with denominator 10.

2. When converting a large number into scientific notation, what is the primary step to determine the exponent?

    (a) Count the total number of digits in the number.

    (b) Move the decimal point so that only one nonzero digit is to the left, and count the number of places moved as a positive exponent.

(c) Move the decimal point to the right and then count the number of zeros.

(d) Count the number of nonzero digits.

3. Which set correctly describes the required range for the coefficient in scientific notation?

   (a) Greater than 0 and less than 1.

   (b) Equal to or greater than 1 and less than 10.

   (c) Greater than 1 and less than 100.

   (d) Any nonzero number.

4. What is $6,500,000$ written in scientific notation?

   (a) 6.5×10^6

   (b) 6.5×10^7

   (c) 65×10^5

   (d) 0.65×10^7

5. What is 0.00042 written in scientific notation?

   (a) 4.2×10^{-3}

   (b) 4.2×10^{-4}

   (c) 42×10^{-5}

   (d) 0.42×10^{-3}

6. In the expression 9.2×10^9, what does the exponent 9 indicate?

   (a) The decimal point in the original number is moved 9 places to the right.

   (b) The decimal point in the original number is moved 9 places to the left.

   (c) The original number contains 9 digits.

   (d) The coefficient 9.2 is multiplied by 9.

7. Which of the following is NOT a step in converting a number to scientific notation?

   (a) Identifying the first nonzero digit.

(b) Moving the decimal point so that the coefficient is between 1 and 10.

(c) Counting the number of digits after the decimal point.

(d) Determining the exponent based on the shift of the decimal point.

Answers:

1. **B:** Scientific notation expresses a number as the product of a decimal coefficient and a power of 10, with the coefficient being at least 1 but less than 10. This method simplifies dealing with very large or very small numbers.

2. **B:** For large numbers, you move the decimal point so that there is only one nonzero digit to its left; the number of places moved becomes the positive exponent on 10.

3. **B:** In proper scientific notation, the coefficient must be equal to or greater than 1 and less than 10.

4. **A:** Converting 6,500,000 involves moving the decimal point 6 places to the left, giving 6.5×10^6.

5. **B:** Converting 0.00042 requires moving the decimal point 4 places to the right to get 4.2, and the exponent is -4, resulting in 4.2×10^{-4}.

6. **B:** The exponent 9 in 9.2×10^9 indicates that the decimal point was moved 9 places to the left to form the coefficient, signifying a very large number.

7. **C:** Counting the number of digits after the decimal point is not part of converting a number into scientific notation. The key steps are identifying the first nonzero digit, moving the decimal point to obtain a coefficient between 1 and 10, and determining the exponent based on the shift.

Practice Problems

1. Convert the large number

$$8\,500\,000$$

into scientific notation.

2. Write the small number

$$0.00076$$

in scientific notation.

3. Express the number in scientific notation

$$3.2 \times 10^5$$

in standard numerical form.

4. Multiply the two numbers in scientific notation:

$$(2.0 \times 10^3) \times (4.5 \times 10^4)$$

and write your answer in scientific notation.

5. Divide the numbers in scientific notation:
$$(6.0 \times 10^7) \div (3.0 \times 10^2)$$
and present your answer in scientific notation.

6. Arrange the following numbers in ascending order:
$$3.5 \times 10^2, \quad 2.8 \times 10^2, \quad 4.2 \times 10^3, \quad 1.1 \times 10^4.$$
Explain your reasoning.

Answers

1. **Solution:**
   To express
   $$8\,500\,000$$
   in scientific notation, we move the decimal point so that only one nonzero digit remains to its left. Placing the decimal after the first nonzero digit (8) and counting the number of moves gives:
   $$8\,500\,000 = 8.5 \times 10^6.$$
   The decimal point has been shifted 6 places to the left, which is why the exponent of 10 is 6.

2. **Solution:**
   For the number
   $$0.00076,$$
   locate the first nonzero digit (7) and move the decimal point to immediately after it. This requires moving the decimal 4 places to the right:
   $$0.00076 = 7.6 \times 10^{-4}.$$
   The negative exponent indicates that the original number is less than 1.

3. **Solution:**
   The scientific notation
   $$3.2 \times 10^5$$
   tells us to shift the decimal 5 places to the right. Doing so gives:
   $$3.2 \times 10^5 = 320\,000.$$
   This is the standard numerical form of the number.

4. **Solution:**
   To multiply the numbers, multiply the coefficients and add the exponents:
   $$2.0 \times 10^3 \times 4.5 \times 10^4 = (2.0 \times 4.5) \times 10^{3+4} = 9.0 \times 10^7.$$
   The result is already in proper scientific notation.

5. **Solution:**
   To perform the division, divide the coefficients and subtract the exponents:
   $$\frac{6.0 \times 10^7}{3.0 \times 10^2} = \frac{6.0}{3.0} \times 10^{7-2} = 2.0 \times 10^5.$$
   Thus, the quotient is expressed in scientific notation as shown.

6. **Solution:**
   Begin by converting each number to its standard form:
   $$3.5 \times 10^2 = 350, \quad 2.8 \times 10^2 = 280,$$
   $$4.2 \times 10^3 = 4200, \quad 1.1 \times 10^4 = 11\,000.$$

Arranging these values from smallest to largest gives:

$$280, \quad 350, \quad 4200, \quad 11\,000.$$

Converting back to scientific notation, the ascending order is:

$$2.8 \times 10^2, \quad 3.5 \times 10^2, \quad 4.2 \times 10^3, \quad 1.1 \times 10^4.$$

The reasoning is as follows: For numbers with the same power of 10, the number with the smaller coefficient is the smaller number. Numbers with a higher power of 10 are automatically larger, regardless of the coefficient.

Chapter 39

Calculations in Scientific Notation

Multiplication in Scientific Notation

Multiplication of numbers expressed in scientific notation is performed by first multiplying the decimal coefficients and then adding the exponents on the powers of 10. This process utilizes the product rule for exponents, which states that for any numbers a and b and any integers m and n,

$$(a \times 10^m) \cdot (b \times 10^n) = (a \cdot b) \times 10^{m+n}.$$

For example, consider the multiplication

$$(2.0 \times 10^3) \cdot (4.5 \times 10^4).$$

The coefficients 2.0 and 4.5 are multiplied to yield 9.0. Next, the exponents 3 and 4 are summed to produce 7. The final expression in scientific notation becomes

$$9.0 \times 10^7.$$

This approach simplifies operations with very large numbers by reducing the calculation to manipulation of a small decimal number and an exponent.

Division in Scientific Notation

Division in scientific notation is accomplished by dividing the decimal coefficients and subtracting the exponent of the divisor from the exponent of the dividend. Mathematically, the division rule is given by
$$\frac{a \times 10^m}{b \times 10^n} = \left(\frac{a}{b}\right) \times 10^{m-n}.$$
As an illustrative example, consider the division
$$\frac{6.0 \times 10^7}{3.0 \times 10^2}.$$
Dividing the coefficients, 6.0 divided by 3.0 results in 2.0. The exponent in the numerator (7) minus the exponent in the denominator (2) yields 5. Thus, the quotient expressed in scientific notation is
$$2.0 \times 10^5.$$
This method streamlines the division process by reducing it to basic arithmetic operations.

Addition and Subtraction in Scientific Notation

Addition and subtraction of numbers written in scientific notation require that the powers of 10 be the same. If the exponents differ, a conversion is necessary so that each term is expressed with an identical exponent before the decimal coefficients can be directly added or subtracted. For instance, consider the sum
$$3.5 \times 10^3 + 2.8 \times 10^2.$$
Since the exponents differ, it is useful to rewrite the second term with an exponent of 3. Recognizing that
$$2.8 \times 10^2 = 0.28 \times 10^3,$$
the addition can now proceed by combining the coefficients:
$$3.5 + 0.28 = 3.78.$$
Thus, the sum is expressed as
$$3.78 \times 10^3.$$

When subtraction is involved, the same procedure applies—ensure that the exponents match before subtracting the coefficients—to maintain uniformity in the expression of the final result.

Normalization of Results

Following arithmetic operations in scientific notation, the resulting coefficient must lie between 1 (inclusive) and 10 (exclusive). In cases where the coefficient does not meet this criterion, normalization, or adjustment, is required. For example, an unnormalized result might appear as
$$15.2 \times 10^4,$$
where the coefficient 15.2 is larger than the acceptable range. By shifting the decimal point one place to the left, the coefficient becomes 1.52 while the exponent is increased by 1 to retain the same value:
$$15.2 \times 10^4 = 1.52 \times 10^5.$$
Similarly, if an arithmetic operation produces a coefficient less than 1, the decimal point is shifted to the right and the exponent is decreased correspondingly. This normalization process ensures that all numbers are expressed in proper scientific notation, thereby preserving clarity and uniformity in calculations.

Multiple Choice Questions

1. Which of the following procedures correctly describes how to multiply two numbers written in scientific notation?

    (a) Multiply the coefficients and multiply the exponents.

    (b) Multiply the coefficients and add the exponents.

    (c) Add the coefficients and multiply the exponents.

    (d) Divide the coefficients and subtract the exponents.

2. In scientific notation, what is the proper procedure for dividing two numbers?

    (a) Multiply the coefficients and subtract the exponents.

    (b) Divide the coefficients and add the exponents.

    (c) Divide the coefficients and subtract the exponents.

(d) Multiply the coefficients and add the exponents.

3. When adding or subtracting two numbers in scientific notation, such as 3.5×10^3 and 2.8×10^2, which step is necessary before performing the operation?

    (a) Convert the coefficients so that they are equal.
    (b) Convert the numbers so that both have the same power of 10.
    (c) Multiply the coefficients and then adjust the exponents.
    (d) Normalize the result after the operation.

4. When the product (or quotient) of an operation in scientific notation yields a coefficient not between 1 (inclusive) and 10 (exclusive), what must be done?

    (a) Multiply the coefficient by 10 and leave the exponent unchanged.
    (b) Normalize the result by adjusting the coefficient and updating the exponent.
    (c) Add a zero to the coefficient without changing the exponent.
    (d) The result is acceptable as is.

5. What is the correct result of multiplying (2.0×10^3) by (4.5×10) in scientific notation?

    (a) 9.0×10
    (b) 9.0×10
    (c) 8.5×10
    (d) 7.5×10

6. Before subtracting 3.5×10^3 and 2.8×10^2, which conversion must be made?

    (a) Convert 3.5×10^3 into 35.0×10^2.
    (b) Convert 2.8×10^2 into 0.28×10^3.
    (c) Convert both numbers into standard decimal form.
    (d) No conversion is needed.

7. An operation in scientific notation produces the unnormalized result 15.2 × 10. What is the normalized form of this number?

   (a) 1.52 × 10
   (b) 1.52 × 10
   (c) 15.2 × 10
   (d) 0.152 × 10

Answers:

1. **B: Multiply the coefficients and add the exponents** This follows from the product rule for exponents: multiply the decimal parts (coefficients) and add the exponents of 10.

2. **C: Divide the coefficients and subtract the exponents** For division in scientific notation, you divide the decimal coefficients and subtract the exponent in the denominator from that in the numerator.

3. **B: Convert the numbers so that both have the same power of 10** Before adding or subtracting, it is essential that both numbers be expressed with the same exponent so that the coefficients can be combined directly.

4. **B: Normalize the result by adjusting the coefficient and updating the exponent** Scientific notation requires the coefficient to be between 1 and 10. When it falls outside this range, you adjust (normalize) by moving the decimal point and modifying the exponent appropriately.

5. **B: 9.0 × 10** Multiplying (2.0×10^3) and (4.5×10) entails multiplying 2.0 by 4.5 to get 9.0 and adding the exponents $(3 + 4)$ to get 7, resulting in 9.0 × 10.

6. **B: Convert 2.8 × 10^2 into 0.28 × 10^3** To subtract these numbers, both must have the same exponent. Converting 2.8 × 10^2 to 0.28 × 10^3 allows you to subtract the coefficients directly.

7. **A: 1.52 × 10** Since the coefficient 15.2 is not between 1 and 10, you shift the decimal one place to the left to get 1.52 and increase the exponent by one, resulting in 1.52 × 10.

Practice Problems

1. Multiply the following numbers in scientific notation:
$$(2.5 \times 10^3) \cdot (4.0 \times 10^5)$$

2. Divide the following numbers in scientific notation:
$$\frac{9.0 \times 10^8}{3.0 \times 10^4}$$

3. Add the following numbers in scientific notation:
$$3.2 \times 10^4 + 4.5 \times 10^3$$

4. Subtract the following numbers in scientific notation:
$$7.8 \times 10^5 - 2.3 \times 10^4$$

5. Normalize the following expression into proper scientific notation:
$$12.4 \times 10^2$$

6. Compute the following multi-step expression and express your final answer in scientific notation:
$$[(2.0 \times 10^3) \cdot (5.0 \times 10^2)] + \left[\frac{1.0 \times 10^6}{2.0 \times 10^0}\right]$$

Answers

1. For the multiplication

$$(2.5 \times 10^3) \cdot (4.0 \times 10^5),$$

first multiply the decimal coefficients:

$$2.5 \times 4.0 = 10.0.$$

Then add the exponents:

$$3 + 5 = 8.$$

This gives an intermediate result of

$$10.0 \times 10^8.$$

Since in proper scientific notation the coefficient must be at least 1 but less than 10, we normalize by writing 10.0 as 1.0 multiplied by 10:

$$10.0 \times 10^8 = 1.0 \times 10^1 \times 10^8 = 1.0 \times 10^9.$$

Therefore, the final answer is:

$$1.0 \times 10^9.$$

2. For the division

$$\frac{9.0 \times 10^8}{3.0 \times 10^4},$$

divide the coefficients:

$$\frac{9.0}{3.0} = 3.0.$$

Subtract the exponent in the denominator from the exponent in the numerator:

$$8 - 4 = 4.$$

Thus, the quotient is:

$$3.0 \times 10^4.$$

This answer is already in proper scientific notation.

3. To add
$$3.2 \times 10^4 + 4.5 \times 10^3,$$
first express both terms with the same power of 10. Rewrite the second term by converting the exponent from 3 to 4:
$$4.5 \times 10^3 = 0.45 \times 10^4.$$
Now add the coefficients:
$$3.2 + 0.45 = 3.65.$$
The sum in scientific notation is:
$$3.65 \times 10^4.$$

4. For the subtraction
$$7.8 \times 10^5 - 2.3 \times 10^4,$$
convert the second term so that both have the same exponent. Rewrite 2.3×10^4 as:
$$2.3 \times 10^4 = 0.23 \times 10^5.$$
Now subtract the coefficients:
$$7.8 - 0.23 = 7.57.$$
Thus, the difference is:
$$7.57 \times 10^5.$$

5. To normalize the expression
$$12.4 \times 10^2,$$
note that the coefficient 12.4 is not between 1 and 10. Divide 12.4 by 10 to get 1.24 and multiply the power of 10 by an additional factor of 10:
$$12.4 \times 10^2 = 1.24 \times 10^1 \times 10^2 = 1.24 \times 10^3.$$
Therefore, the normalized form is:
$$1.24 \times 10^3.$$

6. For the multi-step expression
$$[(2.0 \times 10^3) \cdot (5.0 \times 10^2)] + \left[\frac{1.0 \times 10^6}{2.0 \times 10^0}\right],$$
first solve the multiplication:
$$(2.0 \times 10^3) \cdot (5.0 \times 10^2) = (2.0 \times 5.0) \times 10^{3+2} = 10.0 \times 10^5.$$
Normalize this result:
$$10.0 \times 10^5 = 1.0 \times 10^6.$$
Next, solve the division:
$$\frac{1.0 \times 10^6}{2.0 \times 10^0} = \frac{1.0}{2.0} \times 10^{6-0} = 0.5 \times 10^6.$$
Since 0.5 is below the desired range, normalize it by writing:
$$0.5 \times 10^6 = 5.0 \times 10^5 \quad (\text{because } 0.5 = 5.0 \times 10^{-1}).$$
Now add the two results. To add numbers in scientific notation, they must have the same exponent. Rewrite 5.0×10^5 as:
$$5.0 \times 10^5 = 0.5 \times 10^6.$$
Now add the coefficients:
$$1.0 + 0.5 = 1.5.$$
Hence, the final answer is:
$$1.5 \times 10^6.$$
This answer is in proper scientific notation.

Chapter 40

Problem Solving Strategies in Pre-Algebra

Approaching Mathematical Problems with Logical Reasoning

Pre-algebra problems are designed to develop a structured mindset that emphasizes careful analysis and logical progression. In this context, a problem is viewed as a set of given information accompanied by an unknown that must be determined. The process begins by clearly identifying all relevant quantities and operations that connect them. Logical reasoning involves examining the relationships between elements of a problem, noting patterns, and recognizing the inherent structure. This approach allows intermediate steps to be organized in a deliberate sequence, ensuring that each stage of the solution is based on previously established facts. Often, a statement is dissected into its individual components, where variables and constants are assigned to unknown and known elements respectively, leading to the formulation of an algebraic representation that mirrors the real-world situation.

Developing Systematic Approaches

A systematic approach in pre-algebra emphasizes a step-by-step method to unravel complex problems. The initial phase involves a careful reading of the problem, during which the mathematical elements are identified and cataloged. Converting a word problem into a linear or graphical representation may aid in visualizing the relationships among variables. A systematic framework often includes the following stages: identification of information, organization of data using tables or lists, formulation of appropriate equations, and finally, the execution of arithmetic or algebraic manipulations. By adopting this methodical process, students are encouraged to apply consistency and order to their work, reducing the potential for errors and facilitating the verification of results.

Implementing a Sequential Problem Solving Process

The resolution of a pre-algebra problem is most effective when it follows a defined sequence of steps. First, a thorough examination of the problem statement is essential, ensuring that all necessary details and constraints are understood. Next, the relevant numerical information is extracted and any unknown quantities are designated with appropriate variables. This is followed by the establishment of equations or inequalities that logically connect these quantities. For instance, if the problem involves finding a missing number, one might set up an equation such as

$$a + b = c,$$

where each symbol represents a specific component of the problem. The solution is then derived by performing algebraic operations that systematically isolate the unknown. Throughout this process, attention to detail is paramount, and each algebraic manipulation is performed with careful consideration of the fundamental properties of arithmetic and algebra.

Utilizing Estimation and Verification Techniques

Estimation serves as a valuable tool in assessing the plausibility of an answer, particularly within the framework of pre-algebra where precise calculations are coupled with mental checks. After establishing a numerical solution, estimation techniques can be applied to determine whether the solution falls within a reasonable range based on the given data. Verification, on the other hand, involves retracing the logical steps of the solution to confirm the accuracy of each calculation. A common strategy involves substituting the determined value back into the original equation or problem context. If the equality or relationship holds true, the solution is validated. This practice of verification not only reinforces understanding of the underlying mathematical principles but also cultivates habits that are essential to advanced problem solving.

Illustrative Examples of Strategic Problem Solving

When applying these strategies, several illustrative examples emerge that encapsulate the systematic approach. Consider a scenario where a problem is presented with multiple steps and requires the consolidation of data from different sources. A student might begin by clearly outlining each piece of provided information and then proceed to categorize the data into distinct groups. For example, a problem may require the combination of like terms or the application of a distributive property to simplify an equation. The systematic process could involve writing an intermediate expression such as
$$3x + 2x - 5 = 0,$$
which is then simplified by combining like terms to yield
$$5x - 5 = 0.$$

Subsequently, the equation is solved by isolating the variable. This example demonstrates how a step-by-step strategy, when combined with logical reasoning and careful verification, leads to a reliable and coherent solution process.

Multiple Choice Questions

1. Which of the following is typically the first step in a sequential problem solving process in pre-algebra?

   (a) Identifying all relevant information and unknowns

   (b) Jumping directly into algebraic manipulations

   (c) Estimating the answer without reading the problem

   (d) Checking the solution after completing calculations

2. What is a key benefit of adopting a systematic problem solving approach in pre-algebra?

   (a) It allows you to work hastily without verifying details.

   (b) It minimizes errors by organizing information into logical, manageable steps.

   (c) It replaces the need for logical reasoning with memorization.

   (d) It requires less effort by skipping the analysis of given details.

3. Which of the following techniques is recommended for organizing data when solving a word problem?

   (a) Creating tables or lists to categorize and record the information

   (b) Relying solely on estimating the final answer

   (c) Memorizing similar problems rather than analyzing the current one

   (d) Immediately writing the algebraic equation without understanding the context

4. Why is estimation considered an important tool in the problem solving process?

   (a) It completely replaces the need for exact calculations.

   (b) It helps verify the plausibility of your computed answer.

   (c) It encourages skipping detailed algebraic steps.

   (d) It simplifies the problem by ignoring some of the provided data.

5. Which method best describes how to verify your solution in pre-algebra?

    (a) Rearranging the problem statement and re-reading it
    (b) Substituting your solution back into the original equation or expression
    (c) Comparing your answer with a memorized solution
    (d) Relying solely on estimation to confirm your answer

6. In the illustrative example from the chapter, the expression "3x + 2x - 5 = 0" is simplified to which of the following?

    (a) 6x - 5 = 0
    (b) 5x - 5 = 0
    (c) 3x + 2 = 5
    (d) 5x + 5 = 0

7. Which of the following best embodies the essence of a systematic approach to problem solving in pre-algebra?

    (a) Rushing through the solution to obtain an answer quickly
    (b) Breaking the problem into smaller, manageable steps and verifying each stage
    (c) Relying solely on memorized formulas without analyzing the problem
    (d) Ignoring the relationships between different pieces of information

Answers:

1. **A: Identifying all relevant information and unknowns** This is the first step because a thorough understanding of what is given and what needs to be found lays the foundation for formulating an effective plan to solve the problem.

2. **B: It minimizes errors by organizing information into logical, manageable steps** Adopting a systematic approach ensures that all parts of the problem are addressed in a logical order, reducing the chance of mistakes and enabling a clearer path to the solution.

3. **A: Creating tables or lists to categorize and record the information** Organizing the given data helps in identifying relationships between variables and clarifying the problem's structure before setting up equations or performing calculations.

4. **B: It helps verify the plausibility of your computed answer** Estimation provides a quick check to ensure that the detailed, exact calculations lie within a reasonable range and that the solution makes sense.

5. **B: Substituting your solution back into the original equation or expression** Verifying by substitution confirms that the obtained answer satisfies the original conditions of the problem, ensuring its correctness.

6. **B: 5x - 5 = 0** By combining like terms—adding 3x and 2x—you obtain 5x, which leads to the simplified expression 5x - 5 = 0.

7. **B: Breaking the problem into smaller, manageable steps and verifying each stage** This approach captures the essence of a systematic method by ensuring that each phase of the problem is handled deliberately and methodically, leading to a robust and error-checked solution.

Practice Problems

1. Solve the following word problem using a systematic approach: "The sum of twice a number and 3 is equal to 15." Write an equation that represents the situation and solve for the number.

2. Use a systematic approach to solve this word problem: "In a classroom, there are 5 more girls than boys, and the total number of students is 29. Find the number of boys and girls."

3. Implement a sequential problem solving process to solve the equation:
$$3(x - 4) + 2 = 17$$
Show each step of your work.

4. Apply estimation and verification techniques by doing the following: a) Estimate the product of 49 and 21 by rounding the numbers. b) Compute the exact product, and explain how your estimation compares to the exact result.

5. A student misinterpreted the following problem: "Double a number decreased by 4 equals 10." The student set up the equation as
$$2(x - 4) = 10.$$
Identify the error in this interpretation. Write the correct equation and solve for the number.

6. Describe how you would apply a systematic problem solving strategy to a real-life budgeting problem. In your answer, outline the steps you would take (including identifying variables, organizing data, setting up relationships, solving, and verifying your solution) and explain how each step contributes to finding the solution.

Answers

1. **Solution:** Let the unknown number be represented by the variable x. The phrase "twice a number" translates to $2x$, and "the sum of twice a number and 3" yields the expression $2x + 3$. The problem states that this sum is equal to 15, so we write the equation:
$$2x + 3 = 15.$$

To solve for x, subtract 3 from both sides:
$$2x = 15 - 3,$$
$$2x = 12.$$
Next, divide both sides by 2:
$$x = \frac{12}{2},$$
$$x = 6.$$
Therefore, the number is 6.

2. **Solution:** Let the number of boys be represented by b. Since there are 5 more girls than boys, the number of girls can be expressed as $b + 5$. The total number of students is 29, so we set up the equation:
$$b + (b + 5) = 29.$$
Combine like terms:
$$2b + 5 = 29.$$
Subtract 5 from both sides to isolate the term with b:
$$2b = 29 - 5,$$
$$2b = 24.$$
Divide both sides by 2:
$$b = \frac{24}{2},$$
$$b = 12.$$
Since the number of girls is $b + 5$, we have:
$$12 + 5 = 17.$$
Therefore, there are 12 boys and 17 girls in the classroom.

3. **Solution:** Given the equation:
$$3(x - 4) + 2 = 17,$$

we first apply the distributive property to remove the parentheses:
$$3x - 12 + 2 = 17.$$
Combine like terms ($-12 + 2 = -10$):
$$3x - 10 = 17.$$
Next, add 10 to both sides to isolate the term with x:
$$3x = 17 + 10,$$
$$3x = 27.$$
Finally, divide both sides by 3:
$$x = \frac{27}{3},$$
$$x = 9.$$
Thus, the solution to the equation is $x = 9$.

4. **Solution:** <u>Estimation:</u> To estimate the product of 49 and 21, round each number to a nearby number that is easy to multiply. Rounding 49 to 50 and 21 to 20 gives:
$$50 \times 20 = 1000.$$

<u>Exact Calculation:</u> Compute the exact product: One method is to use the distributive property:
$$49 \times 21 = 49 \times (20 + 1) = 49 \times 20 + 49 \times 1.$$
Calculate each term:
$$49 \times 20 = 980,$$
$$49 \times 1 = 49.$$
Add the results:
$$980 + 49 = 1029.$$

<u>Verification:</u> The estimated product was 1000, which is close to the exact product of 1029. This shows that our estimation is reasonable and provides a quick check on the accuracy of our calculation.

5. **Solution:** The phrase "Double a number decreased by 4 equals 10" should be interpreted as follows: first, double the number ($2x$); then decrease the result by 4, leading to the expression $2x - 4$. Hence, the correct equation is:
$$2x - 4 = 10.$$

The error in the student's interpretation was setting up the equation as
$$2(x - 4) = 10,$$
which would mean subtracting 4 from the number first and then doubling the result. To solve the correct equation, add 4 to both sides:
$$2x = 10 + 4,$$
$$2x = 14.$$
Then divide both sides by 2:
$$x = \frac{14}{2},$$
$$x = 7.$$
Therefore, the number is 7.

6. **Solution:** To apply a systematic problem solving strategy to a real-life budgeting problem, follow these steps:

   (a) **Identify the Problem:** Determine what you need to solve. For a budgeting problem, this might involve balancing income and expenses or finding out how much you can save.

   (b) **Organize the Given Information:** List all sources of income and all expenses. Creating a table or a list can help you keep track of each item.

   (c) **Assign Variables:** Designate variables for the unknown quantities. For example, let S represent the amount you plan to save, I represent your total income, and E denote your total expenses.

   (d) **Set Up Equations:** Use the relationships given in the problem to form equations. For instance, if you want your savings to be the difference between income and expenses, you can write:
   $$I - E = S.$$

(e) **Solve the Equation(s):** Apply algebraic techniques to solve for the unknown variable(s). This might involve combining like terms, isolating a variable, or checking for errors in computation.

(f) **Verify the Solution:** Substitute your answer back into the original equation(s) to ensure consistency and that all conditions of the problem are met. Also, assess if the result is reasonable given the real-life context.

By following these steps, you systematically break down the budgeting problem into manageable parts and increase the likelihood of obtaining a correct and practical solution.

Chapter 41

Analyzing Patterns and Sequences

Foundations of Numeric Patterns

A numeric pattern is an arrangement of numbers in a specific order that follows a determined rule. In these sequences, each term is generated by applying a mathematical operation or a combination of operations to preceding terms or to a fixed initial value. The characteristics of these patterns are embedded in the relationship between consecutive numbers, and the structure may emerge through addition, subtraction, multiplication, or other operations. The study of patterns provides a framework for understanding how numbers relate to one another in organized ways. By exploring the inherent order within these sequences, the regularities become evident even when the numbers appear varied at first glance.

Systematic Identification of Patterns

Identifying the rule of a sequence requires a methodical approach. A common strategy is to calculate the differences between consecutive terms. When the difference remains constant, an arithmetic pattern is present. Alternatively, when a constant factor relates consecutive terms, a geometric pattern is often indicated. In some cases, secondary operations such as examining the ratio or employing multiple steps to find the change reveal more complex behavior.

Each method of identification operates by considering the relationships among the digits, and by evaluating how each term is derived from previous ones. This process involves noting repeated computations or observations that lead to the formulation of a general rule.

Techniques for Extending Sequences

Once a rule governing a sequence is determined, extending the sequence involves applying this rule in a systematic manner to generate additional terms. In an arithmetic sequence, for example, a known first term combined with a common difference permits the use of a formula such as

$$a_n = a_1 + (n-1)d,$$

where a_1 represents the initial term, d the constant difference, and n the position of the term in the sequence. In the case of a geometric sequence, the subsequent term is produced by multiplying the previous term by a constant ratio. The formula

$$a_n = a_1 \cdot r^{n-1}$$

provides a systematic way to determine any term, where r denotes the common ratio. These techniques allow for precise computation of terms without the necessity of a graphical representation, relying entirely on the analytic properties inherent in the numeric relationships.

Worked Examples and Detailed Analysis

A detailed examination of sequences can be accomplished through illustrative examples. Consider an arithmetic sequence where the first term is 4 and the common difference is 3. The sequence begins as 4, 7, 10, 13, and so on. The constant addition operation between successive terms confirms the arithmetic nature of the sequence, and the application of the formula

$$a_n = 4 + (n-1) \times 3$$

allows for computation of any term directly. In a geometric sequence, assume a first term of 2 and a common ratio of 5. The

sequence then unfolds as 2, 10, 50, 250, etc. Verifying the ratio between any two successive terms confirms the multiplicative rule, and the formula
$$a_n = 2 \cdot 5^{n-1}$$
provides the general term accurately. Other sequences, such as those based on squares or triangular numbers, reveal their patterns by recognizing operations such as squaring an integer or summing an increasing series of consecutive integers. The analysis of each example emphasizes the process of discerning the underlying rule through the consistent behavior exhibited by the terms.

Verification of Identified Patterns

A crucial aspect of extending any numeric sequence is verifying that the established rule applies uniformly to all provided terms. Verification involves substituting the term positions into the derived formula and confirming that the computed values match the known terms. In an arithmetic sequence, calculating the difference between each pair of consecutive terms ensures that the differential remains unchanged throughout. Similarly, in a geometric sequence, evaluating the quotient between successive terms confirms the constant ratio. This systematic process of validation solidifies the accuracy of the rule, thereby guaranteeing that the extension of the sequence through the identified method will yield correct and consistent results.

Multiple Choice Questions

1. What best describes a numeric pattern?

    (a) A random set of numbers with no connection.

    (b) A series of numbers chosen arbitrarily.

    (c) An arrangement of numbers following a specific rule.

    (d) A collection of numbers that includes only whole numbers.

2. Which method is typically used to identify an arithmetic sequence?

    (a) Calculating the ratio between consecutive terms.

- (b) Calculating the differences between successive terms.
- (c) Multiplying the terms to see if a pattern exists.
- (d) Arranging the numbers in descending order.

3. What is the formula for the nth term of an arithmetic sequence with first term a_1 and common difference d?
   - (a) $a_n = a_1 + (n-1)d$
   - (b) $a_n = a_1 \times (n-1)d$
   - (c) $a_n = a_1 + nd$
   - (d) $a_n = a_1 \times n + d$

4. Which formula correctly represents the nth term of a geometric sequence with first term a_1 and common ratio r?
   - (a) $a_n = a_1 + (n-1)r$
   - (b) $a_n = a_1 \cdot r^{n-1}$
   - (c) $a_n = a_1 + r^{n-1}$
   - (d) $a_n = a_1 \cdot (n-1)r$

5. What is an essential step in verifying an identified pattern rule in a sequence?
   - (a) Substituting known term positions into the rule to see if the results match.
   - (b) Changing the rule until the numbers appear larger.
   - (c) Ignoring initial terms and focusing only on later ones.
   - (d) Using estimation instead of calculation.

6. Why is it important to determine the pattern rule before extending a sequence?
   - (a) It provides a reliable method to generate subsequent terms.
   - (b) It makes guessing the next number easier.
   - (c) It allows for creative, unpredictable results.
   - (d) It removes the need to calculate any values.

7. Which characteristic distinguishes a geometric sequence?
   - (a) A constant difference between consecutive terms.

(b) A constant addition to each term.

(c) A constant ratio between consecutive terms.

(d) A steadily increasing number of digits.

Answers:

1. **C: An arrangement of numbers following a specific rule**
   A numeric pattern is defined by the rule or relationship that connects one term to the next, rather than being random or arbitrary.

2. **B: Calculating the differences between successive terms**
   In an arithmetic sequence, the key characteristic is a constant difference between terms. Calculating these differences helps verify the arithmetic pattern.

3. **A:** $a_n = a_1 + (n-1)d$
   This is the standard formula for arithmetic sequences. It shows that each term is obtained by adding the common difference d a total of $(n-1)$ times to the first term a_1.

4. **B:** $a_n = a_1 \cdot r^{n-1}$
   In a geometric sequence, every term is generated by multiplying the previous term by a constant ratio r. The formula provided accurately represents this relationship.

5. **A: Substituting known term positions into the rule to see if the results match**
   Verifying the rule means checking that the formula produces the correct known terms in the sequence, ensuring its consistency before extending the pattern.

6. **A: It provides a reliable method to generate subsequent terms**
   Knowing the exact rule behind a sequence eliminates guessing and ensures that any new term generated follows the same predictable pattern.

7. **C: A constant ratio between consecutive terms**
   A geometric sequence is characterized by the fact that the ratio between any two consecutive terms remains constant, distinguishing it clearly from an arithmetic sequence.

Practice Problems

1. Determine the common difference and write the formula for the nth term of the following arithmetic sequence:

   $$7,\ 10,\ 13,\ 16,\ \ldots$$

2. Identify the common ratio and write the formula for the nth term of the geometric sequence:

   $$2,\ 6,\ 18,\ 54,\ \ldots$$

3. Find the 10th term of the arithmetic sequence that begins with 3 and has a common difference of 5. Use the appropriate formula to show your work.

4. Consider the sequence of perfect squares:

$$1, 4, 9, 16, \ldots$$

Write an expression for the nth term of this sequence and calculate the 7th term.

5. Determine if the following sequence is arithmetic, geometric, or neither:
$$5, 5, 5, 5, \ldots$$
Provide a brief explanation for your answer.

6. A geometric sequence starts with 10, and each term is one-half of the previous term. Write the formula for the nth term of this sequence and compute its 5th term.

Answers

1. **Solution:** To find the common difference, subtract the first term from the second term:
$$10 - 7 = 3.$$
Verifying with the subsequent terms:
$$13 - 10 = 3 \quad \text{and} \quad 16 - 13 = 3.$$
The formula for the nth term of an arithmetic sequence is:
$$a_n = a_1 + (n-1)d,$$
where a_1 is the first term and d is the common difference. Here, $a_1 = 7$ and $d = 3$. Therefore, the nth term is:
$$a_n = 7 + (n-1) \times 3.$$
This means that every term in the sequence is 3 more than the previous term.

2. **Solution:** In a geometric sequence, the common ratio r is found by dividing any term by the preceding term. For the sequence:
$$2, \ 6, \ 18, \ 54, \ \ldots$$
We have:
$$\frac{6}{2} = 3, \quad \frac{18}{6} = 3, \quad \frac{54}{18} = 3.$$
Thus, $r = 3$. The formula for the nth term of a geometric sequence is:
$$a_n = a_1 \cdot r^{n-1}.$$
With $a_1 = 2$ and $r = 3$, the formula becomes:
$$a_n = 2 \cdot 3^{n-1}.$$
This formula shows that each term is obtained by multiplying the previous term by 3.

3. **Solution:** For an arithmetic sequence, the nth term is given by:
$$a_n = a_1 + (n-1)d.$$

Here, $a_1 = 3$ and the common difference $d = 5$. Thus, the 10th term is:

$$a_{10} = 3 + (10 - 1) \times 5 = 3 + 9 \times 5.$$

Calculating further:

$$9 \times 5 = 45, \quad \text{so} \quad a_{10} = 3 + 45 = 48.$$

The 10th term of the sequence is 48.

4. **Solution:** The sequence representing perfect squares is:

$$1, \ 4, \ 9, \ 16, \ \ldots$$

Notice that:

$$1 = 1^2, \quad 4 = 2^2, \quad 9 = 3^2, \quad 16 = 4^2.$$

This suggests that the pattern follows the formula:

$$a_n = n^2.$$

To determine the 7th term:

$$a_7 = 7^2 = 49.$$

Therefore, the 7th term of the sequence is 49.

5. **Solution:** Consider the constant sequence:

$$5, \ 5, \ 5, \ 5, \ \ldots$$

In an arithmetic sequence, the common difference is calculated as:

$$5 - 5 = 0.$$

Since the difference is constant (0), it qualifies as an arithmetic sequence. For a geometric sequence, the common ratio is:

$$\frac{5}{5} = 1.$$

Because this ratio is constant, the sequence also meets the criteria for a geometric sequence. Hence, the sequence is both arithmetic (with $d = 0$) and geometric (with $r = 1$).

6. **Solution:** The given geometric sequence starts with $a_1 = 10$ and each term is half of the previous term. This means the common ratio is:
$$r = \frac{1}{2}.$$
The formula for the nth term of a geometric sequence is:
$$a_n = a_1 \cdot \left(\frac{1}{2}\right)^{n-1}.$$
To compute the 5th term, substitute $n = 5$:
$$a_5 = 10 \cdot \left(\frac{1}{2}\right)^{5-1} = 10 \cdot \left(\frac{1}{2}\right)^4.$$
Since:
$$\left(\frac{1}{2}\right)^4 = \frac{1}{16},$$
it follows that:
$$a_5 = 10 \times \frac{1}{16} = \frac{10}{16} = \frac{5}{8}.$$
Thus, the 5th term of the sequence is $\frac{5}{8}$.

Chapter 42

Working with Formulas and Word Problems

Understanding Mathematical Formulas in Real-Life Contexts

Mathematical formulas serve as concise representations of relationships between quantities encountered in everyday scenarios. A formula encapsulates a rule that connects variables and constants in a precise manner. In many contexts, a formula models a situation where numerical values interact in predictable ways. For example, the area of a rectangle is expressed as

$$A = l \times w,$$

where the symbol l denotes the length and w represents the width. This expression provides a clear and consistent method to calculate area regardless of the specific measurements involved. In real-life applications, such formulas help quantify relationships such as cost versus quantity, speed versus time, or volume versus dimensions. The structure of these formulas embodies the ideas developed through pre-algebra skills and lays the foundation for interpreting and solving practical problems.

Translating Real-Life Scenarios into Mathematical Expressions

In many real-life situations, problems present themselves through narrative descriptions. The key to addressing these problems lies in the ability to translate the given situation into a mathematical expression or equation. This process involves identifying the important quantities, assigning variables to unknown values, and writing an expression that captures the relationships among the quantities.

1 Identifying Key Quantities and Variables

The first step in expressing a word problem mathematically involves careful analysis of the information presented. Critical quantities must be identified and classified as either variables or constants. Variables may represent quantities subject to change, such as the number of objects, distance traveled, or time elapsed. Constants, on the other hand, are fixed values such as rates, fixed fees, or known measurements. For instance, in a scenario where the total cost is determined by a fixed service charge in addition to a per-unit price, the constant service charge remains unchanged while the per-unit price is multiplied by the variable representing quantity. The assignment of appropriate symbols for these elements is essential to forming a coherent mathematical representation.

2 Formulating Equations from Contextual Descriptions

Once the key quantities have been identified, the next step is to establish the relationships among them. This typically leads to the formulation of an equation that models the situation accurately. For example, if a problem states that the total cost C of purchasing several items is the sum of a fixed fee F and a cost per item p multiplied by the number of items n, the relationship is mathematically expressed as

$$C = F + p \times n.$$

This equation mirrors the narrative described by assigning each part of the word problem its corresponding mathematical component. The approach relies on pre-algebra skills, such as operations with numbers and variables, to capture the essence of the word problem in a format that permits systematic analysis.

Approaches to Solving Word Problems Through Formulas

Once an appropriate formula or equation has been established from a narrative description, a structured procedure follows for solving the problem. The process involves analyzing the problem statement, isolating the unknown quantity, performing arithmetic or algebraic manipulations, and checking that the solution is consistent with the situation described.

1 Analyzing the Given Problem Statement

A careful reading of the problem statement is critical to discern all given numerical information and identify what is required. The information provided is compared with the expression that has been formulated. For instance, if a problem involves calculating the distance traveled by an object moving at a constant speed, the relationship

$$d = s \times t$$

may be used, where d represents distance, s stands for speed, and t denotes time. A systematic analysis confirms that each variable in the equation is accounted for by a corresponding piece of information in the problem narrative.

2 Implementing Algebraic Techniques to Solve for Unknowns

Once the expression is set up, algebraic techniques are employed to isolate and solve for the unknown variable. This may involve operations such as addition, subtraction, multiplication, or division. In an example where the equation is

$$C = F + p \times n,$$

and both the total cost C and the fixed fee F are known while n remains unknown, the first algebraic step is to subtract the fixed fee from the total cost, giving

$$C - F = p \times n.$$

Subsequent division by the per-item price p then isolates n:

$$n = \frac{C - F}{p}.$$

This series of steps illustrates the methodical approach required in employing pre-algebra techniques to derive a solution that accurately reflects the intended relationship described in the problem.

3 Verifying Solutions with Contextual Consistency

Final verification involves substituting the computed value of the unknown back into the original equation to ensure that the solution satisfies the conditions of the word problem. This step is crucial in confirming that the interpretation of the narrative and the subsequent algebraic manipulations have produced a valid result. When each part of the equation aligns with the contextual information, confidence is established in both the process and the outcome. This verification reinforces the importance of clear correspondence between the mathematical expression and the real-life scenario it models.

Multiple Choice Questions

1. Which of the following best describes the role of a mathematical formula in real-life contexts?

    (a) It provides a random set of numbers for practice.

    (b) It represents a concise relationship between quantities found in everyday situations.

    (c) It is used only in advanced mathematics and is not applicable to daily life.

    (d) It is designed solely to complicate problem solving.

2. What is the first step when translating a word problem into a mathematical expression?

    (a) Guessing the answer and then writing an equation.

    (b) Identifying the key quantities and assigning variables to them.

(c) Writing down an equation without analyzing the problem details.

(d) Calculating the answer using a calculator before forming the equation.

3. In the cost equation
$$C = F + p \times n,$$
which of the following statements is true?

(a) The term F represents a variable fee that changes with every purchase.

(b) The term p is the fixed fee regardless of the number of items.

(c) F is a constant fixed fee, while p represents the per-item cost.

(d) n is the fixed number of items purchased in every situation.

4. Which method is the most appropriate when solving for an unknown variable in an equation derived from a word problem?

(a) Use algebraic manipulation such as addition, subtraction, multiplication, or division to isolate the variable.

(b) Multiply all terms by an arbitrary number to get rid of fractions.

(c) Substitute random values until one fits the scenario.

(d) Rearrange the equation without considering the operations involved.

5. Why is it important to verify your answer by substituting the solution back into the original equation?

(a) It helps confirm that the computed solution satisfies the conditions described in the problem.

(b) It makes the problem-solving process more time consuming.

(c) It is an optional step that can often be skipped.

(d) It introduces unnecessary complexity into straightforward calculations.

6. Which of the following equations correctly models the total cost C of purchasing n items at a cost of p each with an added fixed fee F?

    (a) $C = n + p + F$
    (b) $C = p \times n - F$
    (c) $C = F + p \times n$
    (d) $C = F \times n + p$

7. In the equation
$$d = s \times t,$$
used to calculate distance, what does the variable t represent?

    (a) Temperature during travel.
    (b) Time or the duration of travel.
    (c) The thickness of the object in motion.
    (d) The type of speed measured.

Answers:

1. **B: It represents a concise relationship between quantities found in everyday situations.**
   Explanation: A formula like this encapsulates the rule connecting different measurements and helps translate real-life situations into a mathematical format.

2. **B: Identifying the key quantities and assigning variables to them.**
   Explanation: Before forming an equation, it is essential to understand which numbers are constant and which vary, so assigning variables lays the foundation for accurate problem solving.

3. **C: F is a constant fixed fee, while p represents the per-item cost.**
   Explanation: In this cost model, F stands for a fixed charge that does not change with the number of items, and p is multiplied by the variable n (the number of items purchased).

4. **A: Use algebraic manipulation such as addition, subtraction, multiplication, or division to isolate the variable.**

Explanation: Solving for an unknown variable requires applying proper algebraic techniques to isolate the variable on one side of the equation.

5. **A: It helps confirm that the computed solution satisfies the conditions described in the problem.**
Explanation: Verification by substitution ensures that your calculated value is correct and that the original equation holds true in the given context.

6. **C:** $C = F + p \times n$
Explanation: This equation correctly models the scenario by adding the fixed fee F to the product of the per-item cost p and the number of items n.

7. **B: Time or the duration of travel.**
Explanation: In the formula $d = s \times t$, t stands for time, indicating the period over which the speed s is maintained to cover a distance d.

Practice Problems

1. A local carnival ride costs a fixed entry fee of 5 units plus 2 units for each ride. Write an equation for the total cost,

$$C,$$

when a person takes x rides. Then, compute C when a student rides 7 times. Be sure to identify which term represents the fixed fee and which term represents the variable cost.

2. A rectangular garden has a width of w meters and a length that is 4 meters longer than the width. Write an expression for the area,

$$A,$$

of the garden in terms of w. Then, determine the area when $w = 6$ meters.

3. A cell phone company charges a fixed monthly fee of 20 units plus an additional 0.05 units for each minute used. Write an equation for the monthly bill,

$$M,$$

in terms of the number of minutes m used. Then, calculate the bill for 250 minutes of usage.

4. Two cell phone plans are being compared. Plan A charges a fixed fee of 15 units per month plus 0.10 units per minute. Plan B charges 0.25 units per minute with no fixed fee. Write an equation to find the number of minutes m at which both plans cost the same, and solve for m.

5. A taxi service charges a base fare of 3 units plus 1.50 units per mile traveled. Write an expression for the total taxi fare,

$$T,$$

in terms of the number of miles m. Then, determine how many miles can be traveled for a total fare of 18 units.

6. A sports equipment store sets the selling price of a pair of shoes by adding a 40% markup to the manufacturing cost M. Write an expression for the selling price,

$$P,$$

in terms of M. Then, compute P when $M = 50$ units.

Answers

1. **Solution:**
   The total cost C is composed of a fixed fee of 5 plus a variable cost of 2 per ride. The equation is:

   $$C = 5 + 2x.$$

For $x = 7$ rides:

$$C = 5 + 2(7) = 5 + 14 = 19.$$

Here, 5 is the fixed fee (it does not change with the number of rides), and $2x$ is the variable cost (which increases as the number of rides increases).

2. **Solution:**
The length l of the garden is given by:

$$l = w + 4.$$

The area A of a rectangle is:

$$A = \text{length} \times \text{width} = w(w+4) = w^2 + 4w.$$

When $w = 6$ meters:

$$A = 6^2 + 4 \times 6 = 36 + 24 = 60 \text{ square meters.}$$

3. **Solution:**
The monthly bill M can be expressed as the sum of the fixed fee and the charge per minute:

$$M = 20 + 0.05m.$$

For $m = 250$ minutes:

$$M = 20 + 0.05(250) = 20 + 12.5 = 32.5.$$

Thus, the bill for 250 minutes is 32.5 units.

4. **Solution:**
For Plan A, the cost is:

$$\text{Cost}_A = 15 + 0.10m.$$

For Plan B, the cost is:

$$\text{Cost}_B = 0.25m.$$

Setting the two costs equal to find the breakeven point:

$$15 + 0.10m = 0.25m.$$

Subtract $0.10m$ from both sides:
$$15 = 0.15m.$$

Solving for m:
$$m = \frac{15}{0.15} = 100.$$

Therefore, both plans cost the same when 100 minutes are used.

5. **Solution:**
The total taxi fare T is given by the base fare plus the charge per mile:
$$T = 3 + 1.50m.$$

To find the number of miles when $T = 18$ units:
$$18 = 3 + 1.50m.$$

Subtract 3 from both sides:
$$15 = 1.50m.$$

Solving for m:
$$m = \frac{15}{1.50} = 10.$$

Thus, 10 miles can be traveled for 18 units.

6. **Solution:**
The selling price P is the sum of the manufacturing cost M and a 40% markup on M:
$$P = M + 0.40M.$$

This can be factored as:
$$P = 1.40M.$$

When $M = 50$ units:
$$P = 1.40(50) = 70.$$

Thus, the selling price is 70 units.

Made in the USA
Las Vegas, NV
01 April 2025